Contributing Author
Judy Quadhamer

Editor
Ellyn Siskind

Editorial Project Manager
Mara Ellen Guckian

Editor-in-Chief
Sharon Coan, M.S. Ed.

Illustrators
Kevin Barnes
Renee Christine Yates

Cover Artist
Barb Lorseyedi

Art Manager
Kevin Barnes

Art Director
CJae Froshay

Imaging
Craig Gunnell

Product Manager
Phil Garcia

Publishers
Rachelle Cracchiolo, M.S. Ed.
Mary Dupuy Smith, M.S. Ed.

Full Color

Literacy Centers for Science Skills

Pre K–1

Author
Polly Hoffman

Teacher Created Materials, Inc.

6421 Industry Way
Westminster, CA 92683
www.teachercreated.com

ISBN-0-7439-3399-0

©2004 Teacher Created Materials, Inc.

Made in U.S.A.

# Table of Contents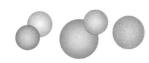

 # Introduction

Science centers encourage students to come and see, come and do, and come and learn. Children approach learning in much the same way that scientists do research—they explore, experiment, and test. Children are full of questions, and science centers encourage them to discover answers through hands-on scientific exploration. Their curiosity will lead them to creative solutions and conclusions.

Literacy Centers for Science Skills includes activities in the following science areas:

- **Nature of Science**—developing the skills of questioning, planning, predicting, implementing, and observing as a way of learning
- **Life Science**—the study of living things, their characteristics, and their relationship to the environment
- **Earth and Space Science**—the study of Earth's elements and structure (rocks, soil, metals, water, air), as well as its place in the universe
- **Physical Science**—the study of matter and its properties (size, shape, color, composition, etc.); the position and motion of objects; the nature of light, heat, electricity, and magnetism

Children want to be involved in their learning. Science provides a way for them to learn about their world and their environment. Lessons in this book are designed to provide a "hands-on" experience so that children can make their own scientific discoveries. Exploration is a goal of each lesson. All the activities in this book incorporate current science standards (see pages 10–11).

Most activities are written for whole-group instruction first, with the centers serving as a follow-up and extension of your introduction. Good teacher preparation and modeling will ensure a successful center experience.

The center approach presents science in a fun, non-threatening manner. Children learn through experience and through discussion. The discussion element is extremely important because it helps cement their acquisition of scientific knowledge. While doing an experiment, children are not only learning new information, they are also developing the vocabulary to go with it. As they participate in the process of an experiment, they begin to learn to ask good scientific questions, examine the world around them more closely, and use what they have observed to predict, form explanations, and reach conclusions. Centers also demonstrate that each student's conclusions may vary, and that this is acceptable.

It is important for teachers to create a learning environment that will enable students to develop a love of science and the sense of "WOW" that goes with it. Each experience that students have with discovering science does more than teach content. Understanding scientific concepts and processes will be a long-range benefit for children growing up in a quickly changing world. It will help them develop the necessary thinking skills to be successful in life.

Getting Started

Getting Started

Most activities or projects listed in this book will take approximately 15–30 minutes. The activities are not necessarily related nor sequential, thus allowing teachers to pick and choose the centers most suited to their students' needs and levels of interest.

Take some time to consider each of the following and check them off as you begin to plan each science center. Use the space at the end to write any helpful notes:

❑ How many students will rotate through the center? Keep in mind that each learning center will be most effective with a limited number of students, sometimes a small group of three or four, sometimes a pair, and sometimes just individuals.

❑ What are the students' ages, abilities, and interests? Will the center accommodate students with special needs?

❑ Do you have enough staff or adult volunteers available for adequate supervision? Welcome parent volunteers into your classroom. Parents are the child's most important teacher and tapping that resource for your classroom benefits the child as well as the teachers. Additional parents or adults in the room provide for small group instruction and enriched vocabulary building. Provide parents with clear instructions, lots of affirmation, and notes of appreciation.

❑ Is enough space available to adequately accommodate both the activity and the number of students participating at any given time?

❑ Can centers be arranged so that they can be easily viewed and supervised?

❑ How will very messy projects be handled? What are the outdoor possibilities for center set up?

❑ Are there ample supplies for each student? Have you provided a rich center environment?

❑ Will there be enough time allotted for students to finish the activity?

❑ Can the topic be extended to other areas of the curriculum? Integrate lessons whenever possible: Young children use math skills such as estimating, measuring, and graphing when recording and evaluating science data. The reading and writing connection is evident through journaling, drawing, dictated writing, and recording data. Overlapping into other subject areas such as literature, music, and art is a fun and effective way to make the study more relevant.

❑ Does the activity build on what students already know?

❑ Does the activity challenge them to apply newly acquired skills elsewhere?

❑ Were your directions clear? Did you spend enough time modeling the activity so that the students will be able to work independently at the center?

❑ Will students succeed?

Notes:

_____

_____

_____

 # How to Set Up a Center

When students work in a cooperative group on a science project, they communicate by sharing ideas, asking questions, drawing, playing, recording, and writing. Activities and centers should be tailored to build these language and communication skills. Science centers can be set up anywhere that space allows. In an optimal situation, there is an existing area in the classroom designated for science materials. The location should be distanced from other students to avoid distractions and disruptions, yet readily visible to allow for supervision.

Outfit the center with related reading materials, small chairs, specimen containers and trays, and writing materials. Add shelves for labeled trays and boxes containing measuring equipment; exploration aids, such as magnifying glasses; tools, springs, wheels, pulleys, etc.; and specimens such as shells, rocks, wood, leaves, nests, eggs, leather, fur, and antlers. Having the supplies readily available allows teachers to capitalize on spontaneous, teachable moments that arise.

Centers should be close to a water supply. If that is not possible, water should be made available in easy-to-carry containers. It is always best to provide coverings for the floor and tables when the projects are messy.

Labeling is as important in the science center as it is in other classroom areas. Signs can be made and displayed above the center area or set on the table. Encourage students to make additional signs for specific center activities, experiments, and specimens. Vocabulary cards should be visible in the center and attached to relate items when applicable.

Keep in mind that materials will vary for each activity. Some may need to be ordered from a catalog, but most are readily available from garden nurseries, grocery stores, or hardware stores. Parents are a good source to help stock your science cupboard. Communicate with them on your specific needs. Talk to local storeowners about donating containers, consumable materials, and other items.

Before conducting any activity with your students, you may want to skim through the entire unit to familiarize yourself with objectives, necessary materials, and time and space constraints. Spend time gathering knowledge about the topic before you work with your students. Read the fact page at the beginning of each section. Visit your school or local library and gather both picture books (stories usually found in the young reader section) and non-fiction books about the topic. Share the books that you have found at a center. Encourage your students to predict what they will be learning about next, and to learn as much as they can before they study it as a class.

Display any posters, pictures, or magazines you have gathered that are related to the topic prior to the actual date the center is introduced. If possible, allow a day or two for students to examine the books and posters and begin formulating questions. Use an enlarged version of the KWHL Chart (page 6) to help facilitate discussions. Motivate students by telling them that they may find out something that you didn't know, and that they will have a chance to be the teacher and share it with the class.

Prior to the day you actually teach the group lesson and open the new center, gather all the materials and resources you will need. Each center's equipment can be stored in a large box with a lid, a clear plastic box, or a similar container. The center can then be easily assembled on a small table or desk when in use.

The better prepared you are, the more success you will have presenting your lesson.

The KWHL Chart is an excellent way to introduce a center to a large group of students and engage them in discussion. A sample of the chart is given below and can be enlarged on chart paper, a white board, or a blackboard, to suit classroom needs.

K	W	H	L
What We Know	What We Want to Know	How We Will Find Out	What We Learned

Science Skills and Goals

The Teacher as Facilitator

- helps students develop a respect for all living things.
- teaches responsibility and care of equipment.
- emphasizes both content and process.
- encourages students to find answers for themselves by asking thought-provoking questions: *What do you think? What did you observe? How can you find out?*
- helps children look for relationships to other concepts or ideas.
- is a careful observer and encourages children to be keen observers: *What does it look like now? What has changed?*
- offers more than one way to record data.
- promotes thinking skills with questions: *What might happen if . . .? Why do you think that happened? Can you predict?*
- sets up a safe, enriching environment and modifies it as needed to meet students' changing needs and/or interests.
- welcomes children's ideas and suggestions and encourages all children to contribute.
- models excitement and interest in the process.
- encourages at-home parent participation for lesson reinforcement and review.

The Student as Participant

- cares for materials.
- expands upon existing knowledge.
- improves scientific vocabulary.
- improves social interaction (communication) as a responsible member of small learning groups (sharing, taking turns).
- increases self-confidence, successfully performing independent activities.
- learns to carry out defined responsibilities within the group. (Properly set-up centers teach children to follow specific instructions to reach a result.)
- makes inferences and predictions based on observations.
- practices listening skills.
- practices positive ways to give feedback to others.
- problem-solves in various ways and with a variety of materials.
- develops a positive attitude toward science and the art of inquiry.

 # How to Use This Book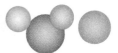

The activities in *Literacy Centers for Science Skills* provide a fun and safe environment for all types and levels of learners. Lessons should be modified to meet the needs of the students involved. Prior knowledge and experiences will need to be taken into account. Materials needed to teach the lesson are listed along with a detailed lesson plan. Each unit includes the following:

The Facts Card: The center activities in this book are meant to be introduced first by the teacher. The Facts Card provides enough scientific information so that the teacher can lead a lively and informed discussion with the class. If time permits, we encourage you to gather further information on the topic. (Studies have shown that the best way to learn is through active participation, and this applies to teachers as well as to young children. Be willing to learn right along with the students! It is essential that you display excitement and curiosity during your lessons.)

The Activity Card: Each center will include a full-color, step-by-step, illustrated card to guide the student through the activity. The student cards have directions similar to those used in the teacher presentation but the text has been simplified. This card should be laminated and placed in the center for review.

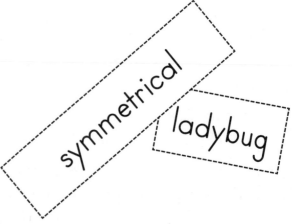

Ladybugs and Symmetry

1. Select a picture.

2. Place the mirror on the dotted line.

3. Look closely at the picture and decide if it is *symmetrical* or *asymmetrical*. Hint: Is the picture created with the mirror the same as the picture by itself?

4. Record your findings.

The Assessment Pages: Student recording sheets, flip-card booklets, and journal pages are just a few of the types of reproducibles included to assist students in gathering information and demonstrating what they have learned. These pages can serve as assessment tools for the teacher in conjunction with classroom discussions. It is important for students to share and discuss their observations and findings. Reviewing each hands-on activity helps cement the learning and increases retention. Displaying students' work on interactive bulletin boards is another way to reinforce their learning. Suggestions are given in each individual lesson.
Assessments will vary with students' ages and the activities.

The Vocabulary Cards: Developing appropriate scientific vocabulary is an important outcome of these lessons. To help facilitate this, vocabulary cards are included for each center. Keep in mind that the purpose of the cards is to expose children to the language and the meaning of the words, not necessarily how to spell them. If appropriate, the vocabulary cards could be placed in the writing and journaling areas or on an interactive bulletin board display for more advanced students. Encourage verbal vocabulary building as well as written usage.

symmetrical

ladybug

How to Use This Book (cont.)

The Teacher Card: The information here guides the teacher in implementing the center. It includes the lesson objective, materials needed, ideas for set-up and presentation, and a home/school connection. You may wish to laminate this page and place it in the center for assistants to use as a reference. Explanations of each section of the teacher card are as follows:

- **The Objective(s)**—Activities in this book are based on National Science Standards and are summarized on page 10. Each center actually meets a variety of standards. (See the checklist on page 11.) Specific objectives are stated at the beginning of each lesson. Planning your science lessons with these standards in mind provides consistency and accountability for students and teachers.

- **Materials**—Materials specific to each center are listed. Read the list carefully to make certain you will have enough time to collect the necessary items prior to presentation. Most items can be found in the classroom, but some may need to be ordered or collected ahead of time.

- **The Lesson**—Introduce the subject prior to beginning the actual center activity to spark student interest. Show photographs from scientific publications or examine a poster. You may want to refer to the first three sections of an enlarged version of the KWHL Chart on page 6 to initiate your presentation. (Later, you will be able to work as a group to fill in the last section of the KWHL chart with what you learned.)

Always state the objective of the lesson. Explain that "cooperation" is the key word in your classroom – both in attitude and behavior. Give the children clear instructions about the expected behavior with specific examples. Practice and role-play these behaviors with young children when centers are first introduced. Discuss the different roles in the group. Remind everyone that even though it is a small group project, each student will do his/her own individual work to record his/her findings. Review recording sheets, mini-books, etc.

Show the class how the illustrated student cards guide them through the activity with step-by-step reminders. Ask for questions or clarifications. Discuss how to take turns and what to do while waiting.

- **Assessment**—Each lesson offers a way in which the student can successfully demonstrate understanding/mastery of the information taught.

- **Home Connection**—This section encourages parent participation with additional activities that reinforce or expand upon what the student has already learned in class.

Science Standards

The chart below displays the McREL standards for science in grades K–2. Used with permission from McREL (copyright 2000 McREL, Mid-continent Research for Education and Learning. Telephone: 303-337-0990. Web site: *www.mcrel.org*).

Earth and Space Science

1. Understands atmospheric processes and the water cycle

2. Understands Earth's composition and structure

3. Understands the composition and structure of the universe and Earth's place within it

Life Science

4. Understands the principles of heredity and related concepts

5. Understands the structure and function of cells and organisms

6. Understands relationships among organisms and their physical environment

7. Understands biological evolution and diversity of life

Physical Science

8. Understands the structure and properties of matter

9. Understands the sources and properties of energy

10. Understands forces and motion

Nature of Science

11. Understands the nature of scientific knowledge

12. Understands the nature of scientific inquiry

13. Understands the scientific enterprise

 # Checking for Standards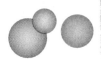

Science Centers	Earth and Space Science	Life Science	Physical Science	Nature of Science
Ladybugs		✓		✓
Spiders		✓		✓
Animals and Their Eggs		✓		✓
Coral Reefs		✓		✓
Shadows	✓		✓	✓
Lakes and Rivers	✓	✓	✓	✓
Stars	✓			✓
Trees and Wood		✓	✓	✓
Water Around Me		✓	✓	✓
Tools and Machines			✓	✓

Parent Letter

Date:_____

Dear Parent(s),

 Your child has been busy at school making scientific discoveries. We are currently learning about

_____.

 We did an activity as a whole group and then took turns exploring further at the class science center. Our objective, based on current science standards, was to

_____

_____.

 Your child is very excited about what he/she has learned and is anxious to share the information with you. Since many experts agree that parent involvement helps children succeed, we would like to encourage you to spend time with your child reviewing and reinforcing what has been learned at school. A copy of the steps we followed for this activity is on the back of this note. Ask your child about each step: "What did you do? What happened? Were you surprised? What did you like about the activity? Can you recite all the steps, in order?"

Enjoy the following family activity to extend your child's science experience:

_____

_____

_____

 Sincerely,

 _____

Ladybugs

Ladybugs are also known as "ladybirds" or "ladybird beetles." There are over 4,000 different varieties of ladybugs and, despite the name, there are male ladybugs.

What are the characteristics of a ladybug? A ladybug is an insect. Like other insects, it has three body parts, six legs, and two antennae. The three body parts are called the head, the thorax, and the abdomen. Like some insects, a ladybug also has wings.

> **Color:** Notice the beautiful color and decorated backs. When we think of ladybugs, we usually think of them as being red with black spots, but they actually come in a variety of colors. Some ladybugs are yellow with black spots, some are black or brown with yellow, white, or red spots, and others are orange with yellow or black spots. The ladybug's color helps it blend in with its environment, especially in a colorful garden. This type of camouflage protects it from insect-eating animals.
>
> **Spots:** The most common ladybug has only two spots, but others have six, ten, or even fourteen spots. Some ladybugs don't have any spots at all.
>
> **Wings:** Ladybugs have two sets of wings, a flying pair and a cover pair. The flying wings are transparent and are tucked under the cover wings when the ladybug isn't using them. The cover wings have a hard surface and protect the flight wings. (See the picture on page 21.)
>
> **Legs:** The legs of an insect are attached to the middle body part, the thorax. Sometimes it looks like the legs of the ladybug are attached to its abdomen because the three body parts are difficult to distinguish on this small insect. Ladybugs are good climbers because they have sticky pads at the end of each of their six legs.

What does a ladybug eat? Ladybugs are beneficial insects because they eat harmful pests in gardens, orchards, and on farms. Aphids, particularly destructive insects, are a ladybug's favorite food. A ladybug can eat 100 of these small insects a day!

What is the life cycle of a ladybug? A ladybug's metamorphosis takes about one month as it transitions from egg to larva to pupa to adult. (See pages 17–18.)

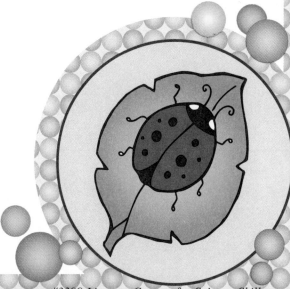

14

Ladybugs

Objectives

Students will use symmetry as a way to develop awareness of the characteristics of organisms; they will learn about life stages and life cycles; they will be motivated to investigate living things and their habitats

Materials

- books about ladybugs and other insects
- activity card, laminated (page 19)
- copies, "Mirror Fun" (page 20)
- enlarged copy of ladybug with wings down (page 21)
- copies, "Same or Different" (page 22)
- flat-edge mirrors (teacher supply store)
- Symmetry Cards, cut and laminated (pages 23–26)
- resealable plastic bag
- Vocabulary Cards, cut and laminated (page 27)

Optional

- ladybugs (local gardening store)
- magazine pictures; puzzles; toy insects; posters
- Life Cycle of the Ladybug, cards cut and laminated (pages 17–18)

Preparation

Arrange fiction and nonfiction books about a variety of insects in a center area. Display pictures, posters, puzzles, and toy insects around the room. Exhibit live ladybugs you might have in a terrarium or other appropriate enclosure. Be sure that you know how to properly take care of them with regard to food, light, and environment. You may want to share this responsibility with your students.

Gather your students on the floor and show them pictures of ladybugs, including a copy of page 21. Ask your students to look very closely at the pictures and describe what they see. Briefly talk about how a ladybug is an insect. Talk about the life cycle of the ladybug and share the picture cards (pages 17–18) with your students.

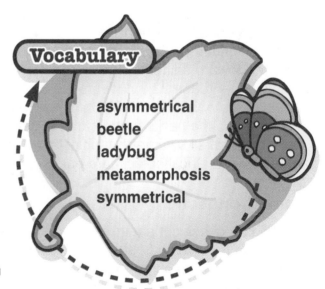

Vocabulary

asymmetrical
beetle
ladybug
metamorphosis
symmetrical

Lesson

Explain to your students that they are going to learn about *symmetry* in nature. Point out that if something is symmetrical, it is exactly the same on both sides of an imaginary line drawn down the middle of the object.

Ladybugs

Lesson *(cont.)*

Show your students an enlarged picture of the ladybug with wings down (page 21). Students will watch you draw a line down the middle of the ladybug. Ask them what they notice about the picture on either side of the line. Now, have them count the spots on each side of the line. Are there the same number of spots on each side? Are the spots in the same place on each side? The body of a ladybug is symmetrical, meaning that it is exactly the same on both sides.

Introduce the science center activity to your students. Tell them that they are going to use mirrors to determine if the object in the picture is *symmetrical*, like the ladybug. Demonstrate how to use a mirror to determine if something is symmetrical. Place the mirror on the line in the middle of the ladybug. Show your students how the image will appear on the mirror. Then, compare the complete image in the mirror with the complete image on the card. (See diagram.) Are they the same? If the answer is "yes," the item is *symmetrical*. If the answer is "no," it is not symmetrical; it is *asymmetrical*.

SYMMETRICAL

Show your students how to follow the activity card (page 19) at the center. Model for your students how to record their findings. Encourage your students to use the vocabulary cards on page 27 as a reference if they will be writing their answers.

As a conclusion to this activity, take a walk and collect objects from nature, including leaves and rocks. Bring them back to your class and see if any of them are symmetrical.

Assessment

There are two types of recording sheets provided to measure student understanding of this activity. The first, on page 20, simply asks students to indicate which objects are symmetrical or asymmetrical. The second, more advanced recording sheet, is geared toward students who are comfortable with writing. On this sheet (page 22), students are asked to write the names of the objects on the Symmetry Cards (pages 23–26). They can use the backs of the cards to self-check their spelling. Also, all symmetrical object cards have purple backs and all asymmetrical object card backs are blue.

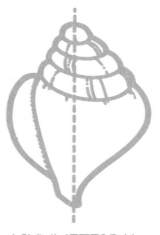

ASYMMETRICAL

Home Connection

Let children check out a mirror to take home and share what they have learned. Encourage them to find new pictures in magazines and books. They may also want to gather objects from home to bring back and share with the class. Ask them to spend time with a parent discussing symmetry.

The Life Cycle of a Ladybug

The process in which an insect develops from an egg into an adult is called *metamorphosis*. Metamorphosis takes about a month. Learning about ladybugs in the spring or summer months is ideal because you can observe them in a local garden, or purchase them from a local garden center to observe in a classroom terrarium. A ladybug's life span is about one year.

Egg

Stage 1

Larva

Stage 2

Pupa

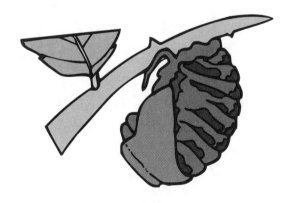

Stage 3

Adult

Stage 4

The Life Cycle of a Ladybug

 ## Stage 2

The larvae are about one third of an inch long (1 cm). They are covered with little sharp, pointy, hairs. The larvae eat other tiny insects, like aphids, and grow quite quickly. They molt (shed their skin) three times within about three weeks.

 ## Stage 1

A ladybug lays between 10 and 100 eggs at a time. The eggs are sticky and yellow or orange in color. They are arranged in small groups under leaves or stems. It takes several days for the eggs to hatch into larvae.

 ## Stage 4

The pupa splits open and out comes a ladybug. When the adult ladybug first comes out, its wings are wet and it has no spots. It is unable to fly for about an hour. Within a few hours, spots begin to appear and the ladybug's color is finalized.

 ## Stage 3

When the larvae are fully grown, they attach themselves to a leaf or branch. They shed their skin once again, exposing the pupa. The pupa is about the size of an adult ladybug, but it is all wrapped up. This protects the ladybug as it develops into an adult. During this stage, the head, legs, and cover wings begin to appear. This is called the pupa stage and will last about one week.

Ladybugs and Symmetry

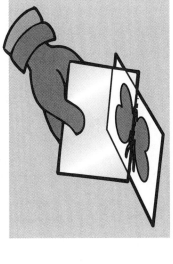

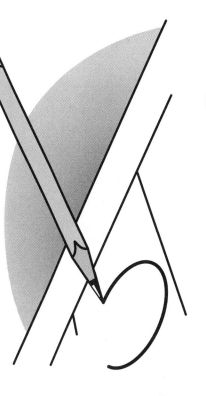

1. Select a picture.

2. Place the mirror on the dotted line.

3. Look closely at the picture and decide if it is symmetrical or asymmetrical. Hint: Is the picture created with the mirror the same as the picture without the mirror?

4. Record your findings.

Mirror Fun

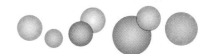

Name: _____

Directions: Look at the different pictures in the boxes below. Circle the objects that are *symmetrical*. Put an X on objects that are *asymmetrical*.

Ladybugs

Same or Different

Name: _____

Symmetrical

Directions: Write the names of the objects that are *symmetrical* on the lines below.

_____   _____
-    -
_____   _____

_____   _____
-    -
_____   _____

_____   _____
-    -
_____   _____

Asymmetrical

Directions: Write the names of the objects that are *asymmetrical* on the lines below.

_____   _____
-    -
_____   _____

_____   _____
-    -
_____   _____

_____   _____
-    -
_____   _____

Symmetry Cards

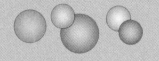 # Symmetry Cards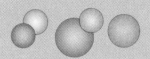

arrow

apple

flower

butterfly

house

heart

airplane

ladybug

 # Symmetry Cards

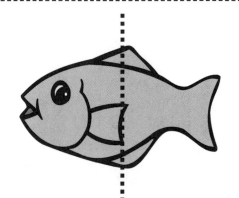

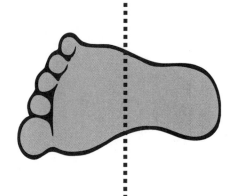

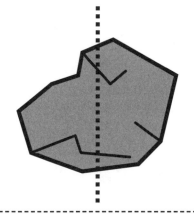

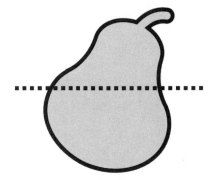

 # Symmetry Cards

fish

tree

foot

cloud

pear

rock

grapes

shoe

 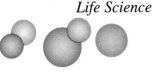
ladybug

beetle

symmetrical

asymmetrical

metamorphosis

 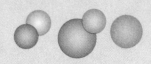
Ladybugs   Ladybugs

Ladybugs

Ladybugs

Ladybugs

Spiders

Is a spider an insect? No. Spiders, along with scorpions, mites, and ticks, are classified as *arachnids*.

What is the difference between an arachnid and an insect? The body of an arachnid is divided into two sections: the c*ephalothorax* (head + thorax) and the *abdomen*. Insects have three distinct body parts: the head, abdomen, and thorax. Arachnids have four pairs of segmented legs attached to the head/thorax, not the abdomen. Insects have only six legs. The spider's six back legs are used for walking. The pair of legs nearest the head is used for holding materials. Spiders do not have antennae or wings while many insects do.

Are all spiders alike? More than 3,000 different species of spiders have been identified in North America alone!

How many eyes does a spider have? Most spiders have two rows of four eyes, or eight eyes in all.

Do spiders lay eggs? Yes! Some spiders lay eggs and attach them to a twig or leaf. Other spiders spin a silk cocoon around their eggs. Still other spiders create an egg sac around the eggs and deposit them on a leaf or wall. Newly hatched spiders are called *spiderlings*.

Why don't spiders fall from the ceiling? Spiders can walk up walls and across ceilings because they have grip-pads on their feet to help them stick.

What does a spider eat? For the most part, spiders eat *invertebrates*—insects and small animals without backbones. (There are, however, some larger spiders, like the tarantula, that can eat small vertebrates like frogs or birds!)

How does a spider catch its food? Some spiders catch their food in webs they have spun. Others run after their prey, physically overpower them, and immobilize them with venom.

Are spiders poisonous? Most spiders are not poisonous, but the bite of a black widow or a brown recluse spider can make a human seriously ill. Spiders are not aggressive and usually use their venom only in self-defense.

How, and why, does a spider make a web? Silk is formed in special glands beneath the spider's abdomen and then squeezed out of two small tubes called *spinnerets*. Once released, the liquid silk dries into a solid strand. Different spiders spin different kinds of webs. Garden spiders build round webs, house spiders build cobwebs, grass spiders build a funnel web in the ground, and poisonous black widows build a formless web. Most spiders spin silken webs to collect food. Spiders do not get stuck in their own webs. Special claws on each foot help them walk without sticking.

Should we be afraid of spiders? No! Spiders are frightened of humans and will run away from them. We should be thankful that spiders are around. They help keep the insect population in check.

Spiders

Objectives

Students will be able to identify spiders by their characteristics and differentiate them from insects. Through discussion and observation, students will develop an appreciation of spiders and their role in the environment.

Materials

- books, pictures, and posters about spiders and their webs (include pages 36–37)
- activity card, laminated (page 33)
- copy, "Parts of a Spider", laminated (page 35)
- chenille sticks
- dough or clay
- felt pens
- glitter, glue
- scissors
- small beads
- spider specimens
- copies, web patterns (pages 38–40)
- Vocabulary Cards, cut and laminated (page 41)

Preparation

Display books, posters, models, available specimens, and pictures of spiders and their webs in the center area. Some children will particularly enjoy looking at copies of "Spider-Man" comic books or the action figures. (*What traits does Spider-Man have that make him special?*)

Lesson

Share books about spiders during group time. Take your students outside for a spider hunt so they may observe spiders (remind them not to touch!). If possible, display spider specimens (in a secure container) that have been collected by adults. Your local pet shop may be willing to loan you a larger spider, or they may have an employee who would be willing to bring one in to share with the class.

Discuss the differences between spiders and insects. Compare the number of body parts, the number of legs, the presence of wings, antennae, etc. Create a chart or Venn diagram illustrating the differences between spiders and insects.

Explain to students that they will be making spiders using clay or dough and chenille sticks. The chenille sticks can be cut to create different-sized legs for the spiders.

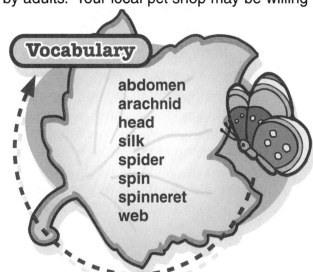

Vocabulary

abdomen
arachnid
head
silk
spider
spin
spinneret
web

Spiders

Lesson (cont.)

Offer a variety of pictures of spiders to look at while working at this center. Include "Parts of the Spider" (page 35). Using clay or dough, model how to make two separate body parts. Connect them with a chenille stick. (Remind students that an insect would have an additional body part called a *thorax*.) Then, count out eight legs for the spider. (Review that an insect has three pairs of legs, not four.) Attach the legs to the head/thorax. Add eyes to the spider's head by poking holes with the end of a chenille stick or with small beads. Make two rows of four eyes each. Finally, talk about different types of webs. Review the web pattern pages and pictures from magazines. Students can trace the web patterns with felt pens, or use glue and glitter to create a raised web.

Assessment

Examine each student's clay spider for two separate body parts and four pair of legs. Are the legs attached to the head/thorax? Can the students verbalize how their spider is different from an insect?

Home Connection

Encourage your students to go on a spider hunt at home with adult supervision. Send home the instructions below for making a temporary spider house for short-term observation.

Temporary Spider Habitat

Materials

- clear jar (any size)
- leaves, grass, twigs, etc.
- waxed paper
- rubber band
- small, moistened sponge
- small stick

Directions

Find a spider. Place the opening of the jar near the spider. Use a small stick or tool to gently nudge the spider into the jar. Be careful — you don't want to hurt the spider! After the spider is safely in the jar, add a few pieces of grass or other items from the area to the jar. Place a moist piece of sponge in the container as a source of water. Cover the top of the jar with the waxed paper. Poke a few small air holes in the paper. Secure it by placing a rubber band around the mouth of the jar. Now, enjoy observing your spider. Keep the spider for a few hours and then be sure to set it free, as it will need to find water and catch food.

Make a Spider

1. Make two body parts for a spider. Connect the two parts with a chenille stick.

2. Add eight legs to the head part.

3. Poke eight small holes for eyes.

4. Trace a web for the spider to rest on.

#3399 Literacy Centers for Science Skills

34

Parts of a Spider

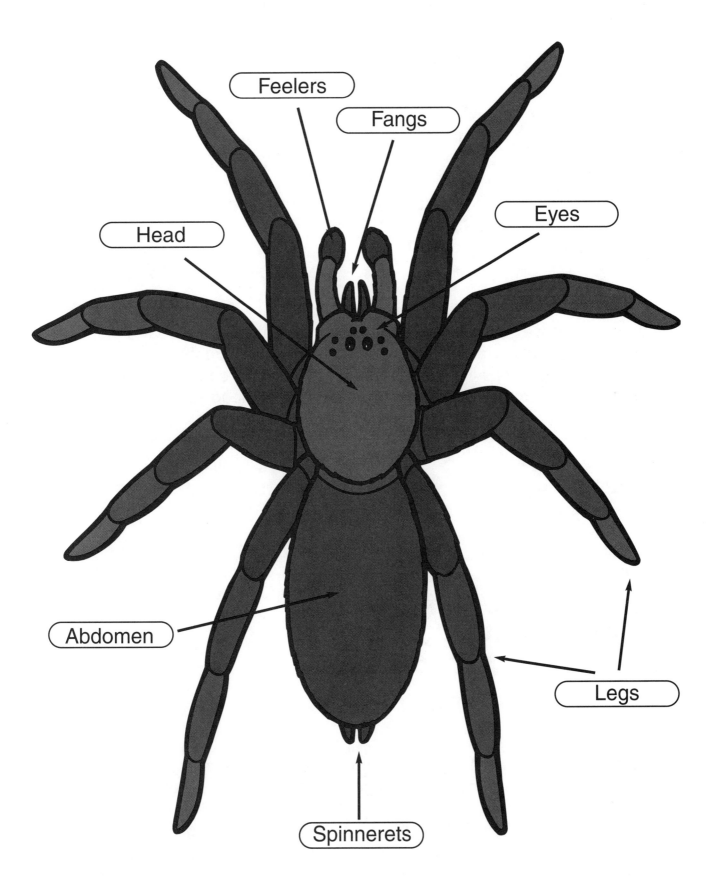

Feelers

Fangs

Eyes

Head

Abdomen

Legs

Spinnerets

Different Spiders

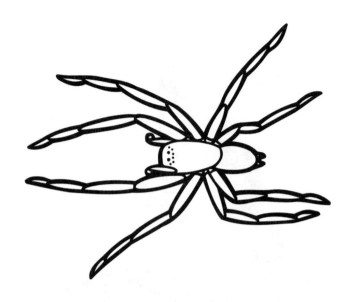

The House Spider

The common house spider is usually found in the bathroom or near a place where it can get water. It is a messy eater. It spits out food that it can't chew. A house spider's web looks like a sheet or hammock. The center, where it catches its prey, is loosely woven. The web is held in place by long strands attached to walls or furniture.

The Orb Weaver

The golden orb weaver gets its name from the color of its web. It spins the strongest web of all spiders. An "orb web" can be six meters high (approximately 20 feet) and two meters wide (approximately seven feet) and can last several years. Often the web is spun between two trees or tree branches. It is designed to catch large insects.

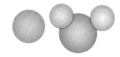
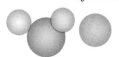

Different Spiders

Crab Spider

The crab spider uses camouflage as a way to trap its prey. It will wait until its food walks right by it. Then, it bites its prey and kills it with its poison. If you tease a crab spider, it will widen its legs and walk sideways like a crab.

Black Widow

The black widow is identified by the red hourglass shape on its underside. It tends to be shy and doesn't like battles. It is poisonous. A bite from a female black widow can kill a human being.

Water Spiders

Water spiders live in or around freshwater lakes and ponds. They live and lay their eggs in underwater bell-shaped webs. They use their legs to swim. They eat midge larvae and water mites. They are eaten by fish and frogs.

Triangle Web

Name: _____

Directions: Trace the web with felt pens or glue. Cover glue with glitter.

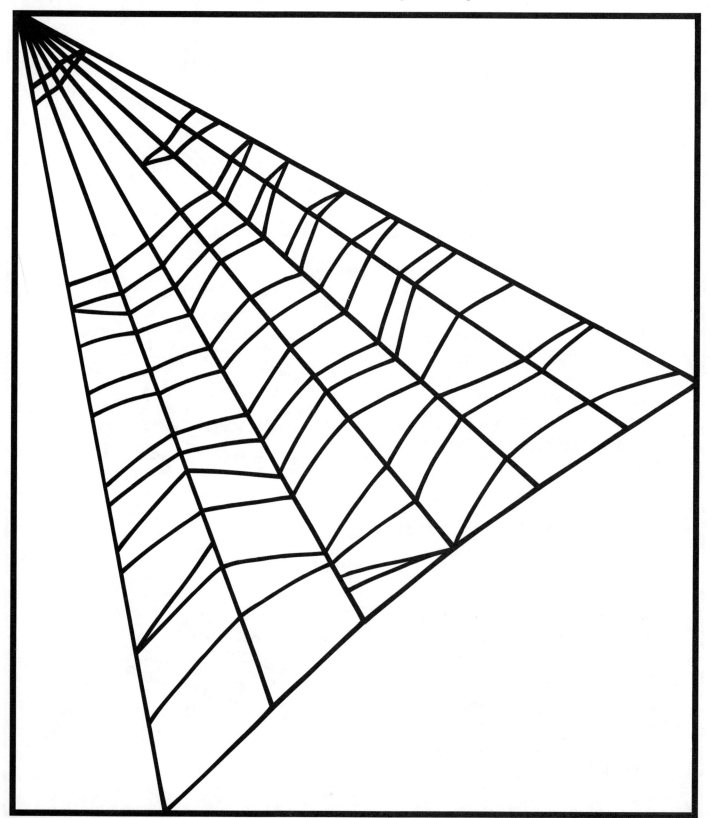

Orb Web

Name: _____

Directions: Trace the web with felt pens or glue. Cover glue with glitter.

Sheet Web

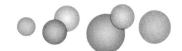

Name: _____

Directions: Trace the web with felt pens or glue. Cover glue with glitter.

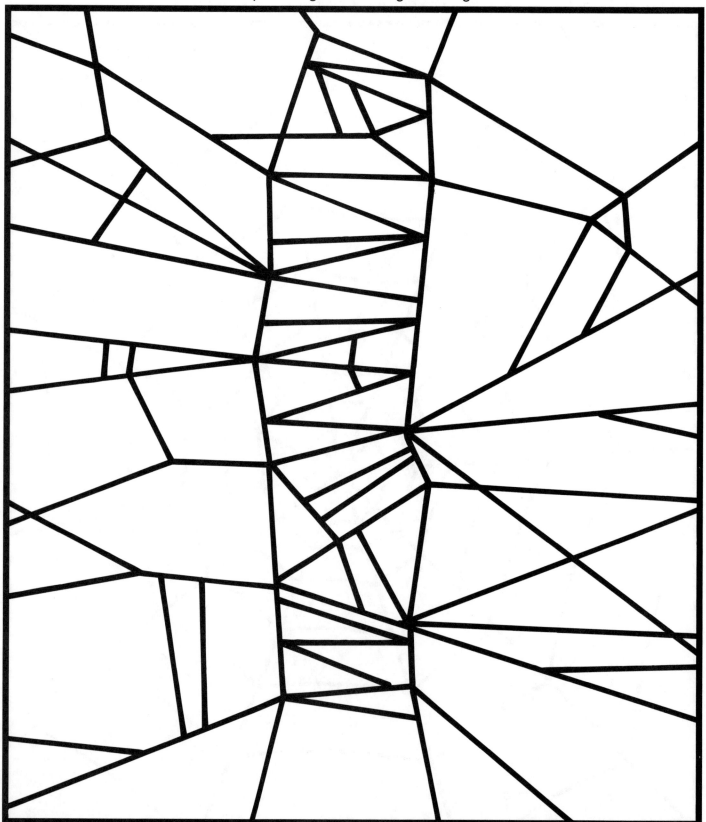

 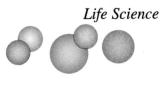
abdomen	arachnid
head	silk
spider	spin
spinneret	web

Vocabulary Cards

Spiders

Spiders

Spiders

Spiders

Spiders

Spiders

Spiders

Spiders

42

Animals and Their Eggs

Do all animals lay eggs? For the most part, birds, reptiles, amphibians, fishes, and insects lay eggs. Characteristics of egg-laying animals include feathers, scales, or slimy skin. Mammals give birth to live young.

What is an egg? An egg is the round-shaped reproductive body produced by the female of certain animal species. It contains everything needed to begin a new animal life.

What are the parts of an egg? An egg consists of an embryo surrounded by nutrient material and covered with a protective shell. There are four main parts to an egg—the *shell*, the *membrane*, the *white,* and the *yolk*. The shell and membrane are the outermost parts of the egg and protect its contents. They keep the egg from drying out, and keep harmful bacteria and dust from entering the embryo. The white of the egg contains *albumin*—a mixture of proteins that protects the yolk and supplies it with food. The yolk, the yellow center of the egg, is the part of the egg that will grow into a new animal.

What color is an egg? Eggs can be many different colors, thus helping to camouflage and protect them from predators. Eggs laid on leaves are often greenish in color, while eggs laid in reeds or low bushes can be beige, brown, or speckled to blend into their environment.

How do eggs hatch? Breaking through an eggshell is hard work. A chick uses a point on the tip of its beak, an "egg-tooth," to crack its shell. The first crack is called a "pip." The chick continues to make pips in a row around the larger end of the egg. When the row is almost complete, the chick gives a huge push and the top of the egg comes off!

Do eggs come in different shapes and sizes? Eggs can be all different shapes, though they are usually round, oval, or pear-shaped. There are some insect eggs, however, that are sharply pointed! The size of the egg depends upon the size of the animal. Hummingbirds, the smallest of the bird family, lay the smallest eggs. Ostriches, the largest of the bird family, lay the largest eggs.

Where do animals lay eggs? Some animals (birds) lay eggs in trees or nests. Some animals (fish, reptiles, and most amphibians) lay eggs in water.

Who cares for the eggs? If an animal lays just a few eggs, each egg is precious and is usually well-cared for. If an animal lays hundreds or thousands of eggs (like most fish in the open sea), they are usually just deposited and left alone. In most cases, the female takes care of the eggs. (The female Emperor penguin, however, lays only one egg and then passes it to the father, who keeps it on top of his feet until it hatches.) Birds diligently care for their eggs, keeping them consistently warm and regularly turning them. On the other hand, sea turtles dig holes in the sand, deposit their eggs, cover them up, and go back into the sea. The sun, shining on the sand, keeps sea turtle eggs warm. Water below the sand keeps the eggs moist.

Animals and Their Eggs

Objectives

Students will learn that some animals produce eggs and some animals produce live offspring. Students will be able to identify examples of each. They will also recognize that offspring resemble their parents.

Materials

- books, pictures, and posters of eggs
- activity card, laminated (page 47)
- eggs posters, laminated (pages 53–56)
- animal cards, laminated (pages 49–50)
- plastic resealable bags
- copy, "Hatched from an Egg", laminated (page 51)
- copies, "Who Lays Eggs?" worksheet (page 59)
- "Chick Development" mini poster (page 57)
- egg specimens and nests
- plastic eggs
- Vocabulary Cards, cut and laminated (page 27)

Preparation

Arrange books about eggs and egg-laying animals in the science center. Display posters and pictures of eggs and egg-laying animals. A local nature center may have nests and/or eggs that they would be willing to lend you. For the truly adventurous, you may even want to look into renting an incubator to hatch your own chicks!

Lesson

Discuss with students what they already know about eggs and egg-laying animals. Ask what else they would like to know and add the information to a KWHL chart (see page 6).

Draw a simple diagram of the four parts of an egg (see page 46). Animals that have feathers, scales, or slippery/slimy skin begin life inside of eggs. Begin a list of animals that hatch from eggs, using the criteria listed. Ask your students to name animals that have feathers, scales, etc. Then, make a second list of animals that give birth to live offspring. Tell your students that these are called *mammals*. Tell them to think of animals with hair/fur.

Share the egg posters (pages 53–56) with the students. Look at the different colors and shapes of eggs. Note that a turtle's egg is round and about the size of a Ping-Pong ball. Hold the egg pictures up to the light so that students can see the animals inside. Can more animals be added to the list after viewing the posters?

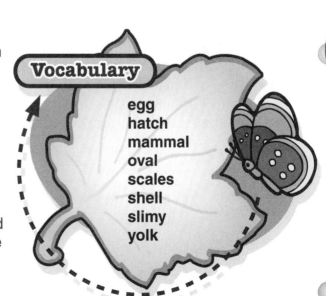

Vocabulary

egg
hatch
mammal
oval
scales
shell
slimy
yolk

Animals and Their Eggs

Lesson *(cont.)*

Refer to the Facts page for information about eggs. Discuss the color and shapes of eggs. Talk about where eggs can be found. Ask your students about the parts of an egg. Discuss how some animals lay numerous eggs at a time, while others lay only a few.

Introduce the animal picture cards (page 49) and "Hatched from an Egg" (page 51). For fun, you might like to place each card in a plastic egg before the presentation. Take turns drawing a card. Talk about the pictured animal and in which category it belongs. For example, a cow would belong in the "Born Alive" category, and a chicken would belong in the "Hatched from an Egg" category. Model for the students how to place the animal cards correctly on the sheet. Tell them that they will have a chance to do this activity on their own in the center.

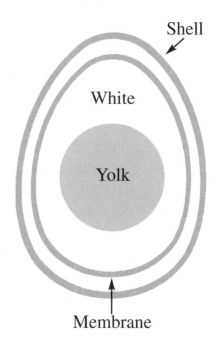

For additional reinforcement, introduce "Who Lays Eggs?" (page 59). Demonstrate how to circle the appropriate animals. You may also wish to share the "Chick Development" mini poster on page 57.

When the activities have been thoroughly modeled and students are comfortable with the procedures, invite them to take turns in the center.

Assessment

After students have sorted all of the animal cards, have them flip the cards over to self-check. There will be an egg on the back of all animals hatched from eggs. Then, check their worksheets to see if they circled the correct animals.

Home Connection

Have the students pick an egg-laying animal to research by taking a trip to their local library to find out all about this animal. Does it lay one egg at a time, a few eggs at a time, or hundreds of eggs at a time? Do the parents take care of the egg? Do they take care of the newborn? For how long? Have them find a picture or draw the animal and its eggs and share the information with the class.

Animals and Their Eggs

dog

frog

chicken

1. Look at the animal cards.

2. Place each animal card under the correct heading.

3. Flip the cards over to check your answer.

4. Complete the worksheet.

Animal Cards

Animal Cards

Hatched
from
an Egg

Born

Alive

Hatched
from
an Egg

Hatched
from
an Egg

Born

Alive

Hatched
from
an Egg

Hatched
from
an Egg

Born

Alive

Hatched
from
an Egg

Hatched
from
an Egg

Born

Alive

Hatched
from
an Egg

Hatched from an Egg

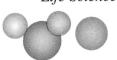

Hatched from an Egg 🥚 🐻	Born Alive 🐻

 # Animals and Their Eggs

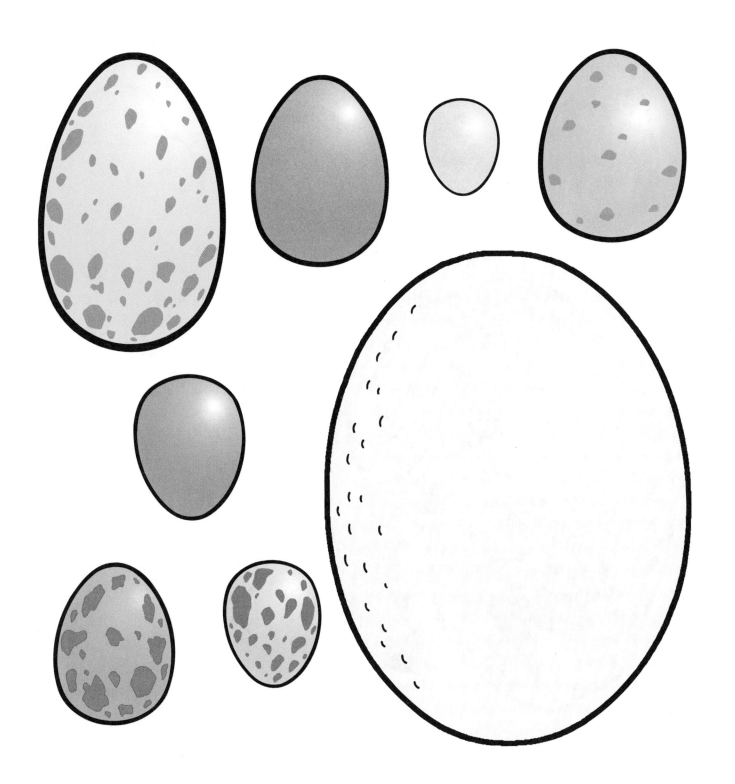

Animals and Their Eggs

 # Animals and Their Eggs

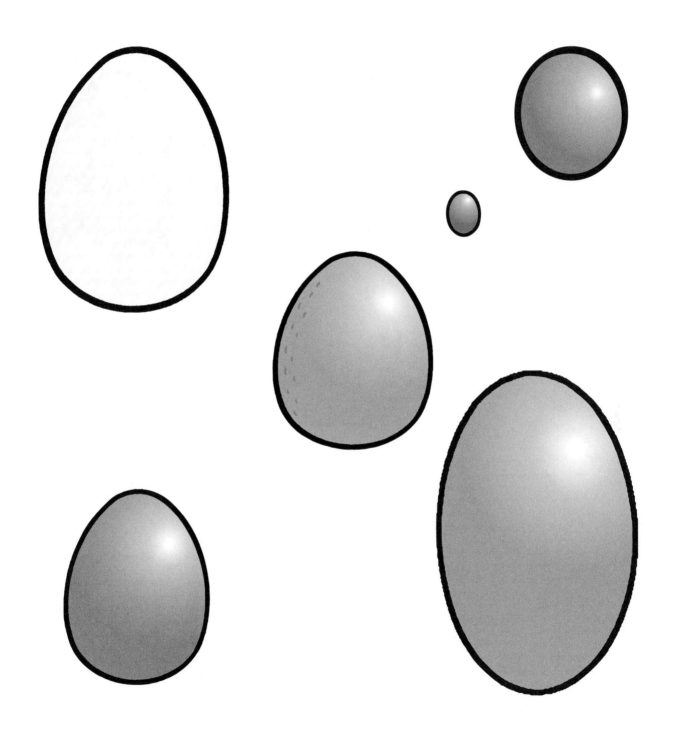

Animals and Their Eggs

Chick Development

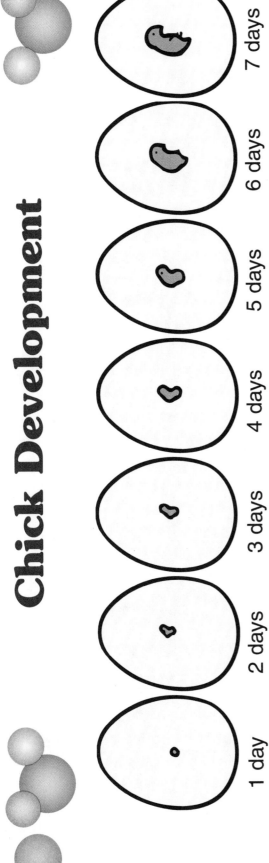

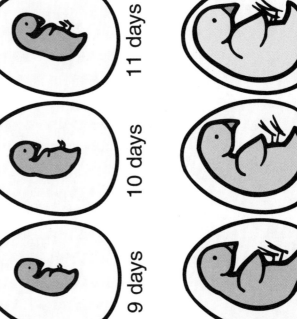

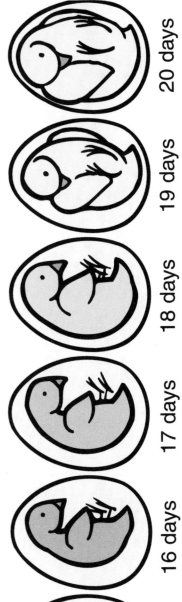

1 day	2 days	3 days	4 days	5 days	6 days	7 days
8 days	9 days	10 days	11 days	12 days	13 days	14 days
15 days	16 days	17 days	18 days	19 days	20 days	21 days

#3399 Literacy Centers for Science Skills

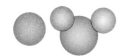

Who Lays Eggs?

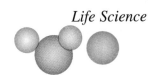

Name: _____

Directions: Color the animals that hatch from eggs.

Life Science

 # Which Came First?

Name: _____

Directions: Look at the pictures below. Number them from 1 to 6 in the order in which they happened.

 # Vocabulary Cards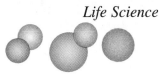

egg	hatch
mammal	oval
scales	shell
slimy	yolk

Vocabulary Cards

Animals & Eggs   Animals & Eggs

Animals & Eggs   Animals & Eggs

Animals & Eggs   Animals & Eggs

Animals & Eggs   Animals & Eggs

Coral Reefs

What is a coral reef? Coral is a not a plant, it is an *invertebrate*—an animal without a backbone. A *coral reef* is an underwater ridge made by a variety of coral, other species, and their skeletons. They often look like exotic forests or gardens. They serve as havens for a wide variety of plant and animal life and protect shores from erosion.

How are coral reefs formed? Coral reefs form in shallow, warm waters. Many different plants and animals, including coral, work together to create reefs. Coral *polyps*, small, tiny animals the size of a pea, huddle together and form colonies. Special cells in the polyp's outer skin create a solid limestone cup that cements the polyp to the reef. As the coral grows, the older parts die off, creating skeletons. New coral builds on the old, creating larger and larger forms. Tubeworms and mollusks add their hard skeletons to the reefs. The formation of coral reefs (animal skeletons) is a slow process.

What and when do coral eat? Coral polyps eat *zooplankton* (newly hatched shrimp, crabs, fish, and sea worms). They use their stinging tentacles to reach out and catch prey that drifts within striking distance. They also eat tiny plants, called *algae*, that live inside their skeletons. Polyps hide during the day and feed at night.

What does coral look like? Coral comes in many different colors and configurations. However, all coral grows in layers, and the color is only on the top layer; the dead layers underneath (skeletons) are white. Species of coral grow differently depending on the amount of sunlight and the force of the waves. Some of the more interesting types of coral are brain coral, elkhorn coral, fan coral, and finger coral. Some coral grow as branches, others as mounds, and others in ridges.

What animals live on a coral reef? Coral reefs are one of the most productive communities on Earth and are often referred to as "underwater cities." One third of all the world's species of fish live in coral reefs. Sharks and sea turtles seek out the warm waters and abundance of food. Reefs provide food and shelter to various types of fish and invertebrates including starfish, parrotfish, sea urchins, giant clams, sea slugs, and clownfish. Other ocean animals, including sponges, worms, and sea urchins, break down the older coral skeletons to create their own homes in the reefs.

Where are coral reefs found? Coral reefs are found in clear, warm ocean waters. The water surrounding reefs must be shallow (6–100 feet deep). If the sea level rises too much and blocks out the sun's rays, the algae will die and the coral will not have food to survive.

Are all reefs the same?

There are three main types of reef formations:

—*Fringing reefs* have coral islands that form a fringe, or edge, along a shoreline.

—*Barrier reefs* run along the shoreline but are separated from it by a lagoon. (Great Barrier Reef, Australia)

—*Atolls* are ring-shaped coral with a lagoon in the center. (Marshall Islands)

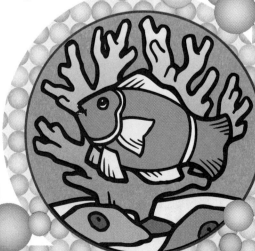

64

Coral Reefs

Objectives

Students will learn what coral is; be able to identify different species of coral and other animals that inhabit a coral reef; understand that distinct environments support the life of different kinds of plants and animals.

Materials

- pictures of coral reefs and their inhabitants
- books about coral reef life
- coral and seashells
- Coral Reef poster, taped together and laminated (pages 69–71)
- mask patterns, cut and laminated (page 73)
- coral and fish cards, cut and laminated (pages 75–77)
- elastic
- hole punch
- clay, different colors
- drinking straws
- chenille sticks
- scissors
- sharpened pencils
- "Types of Coral" laminated mini poster (page 79)
- copies, "Who Lives in the Reef?" (page 80)
- scuba or snorkeling gear
- plastic resealable bags for cards
- Vocabulary Cards, cut and laminated (page 81)

Preparation

Arrange examples of coral and seashells, nonfiction picture books about ocean life, and pictures and posters of coral at the science center. Exhibit scuba and snorkeling gear, if available. If possible, set up a small aquarium in your classroom. Or, go to a pet store—they will have plastic fish and coral for aquariums that you can display at the center.

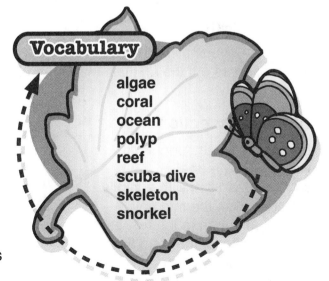

Vocabulary

algae
coral
ocean
polyp
reef
scuba dive
skeleton
snorkel

Lesson

Keep in mind that many students have seen animated movies that include a coral reef. If appropriate, use this as a starting point for discussion about coral reefs. Remind students that the ocean is the largest habitat for both animals and plants on Earth.

Coral Reefs

Lesson *(cont.)*

Read books about coral and coral reefs. Use the chalkboard or a KWHL chart (page 6) to record student input during discussions. Share the posters (pages 69, 71, 79) showing different types of coral and animal life. Encourage students to draw different types of coral. Show examples of coral, if possible. Talk about how coral is formed. Remind students that coral is an animal, not a plant. Explain that coral needs food and sunshine to survive, even though it lives under water.

Demonstrate how to create a coral reef. Use clay to support different layers of coral species: cut snippets of drinking straws (tube coral); roll small balls of clay and etch them, using a pencil point, with thin, squiggly lines (brain coral); curve and connect several short pieces of chenille sticks (finger coral). Can your students see how the coral got its name? Allow time for students to work with the materials and create their own reefs. Refer them to the pictures of different types of coral so that they can use available materials to create other coral species.

Explain the difference between scuba diving and snorkeling. (A scuba diver carries a tank of oxygen for air and can dive far below the water surface for a long period of time. A snorkeler gets air from a tube that connects from his/her mouth to the air above the water's surface.) Ask your students if anyone has been snorkeling or scuba diving and to share their experiences. Share books about snorkeling and scuba diving. Explain that most of our knowledge about the ocean comes from divers. Show pictures of some of the fish that live in a coral reef. Can your students explain how some of the fish got their names? Talk about what it would feel like to be under water and what they would most like to see.

The Coral Reef center will offer students an opportunity to pretend to be divers. Explain that they may wear the mask (page 73) as they try to match the different types of fish and coral on the laminated cards (page 75–77) to their location on the poster.

Optional: Make additional color copies of the cards before laminating them and teach your students how to play "Go Fish!" They will have to learn the names of all the fish and coral in order to ask for them by name as part of the game.

Assessment

Students will complete "Who Lives in the Reef?" (page 80) in which they will cross out the animals that do not inhabit a coral reef.

Home Connection

Encourage your students to visit a pet store that specializes in fish with a family member. Ask them to pay special attention to the tropical tanks and see if they can find any coral skeletons or fish that might live in a reef.

Coral Reefs

1. Lay out the cards around the poster.

2. Put on the mask.

3. Pick a card and find its match in the reef.

4. Match all the cards to the poster.

68

Coral Reef Poster

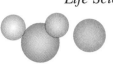

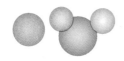

Coral Reef Poster

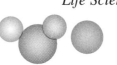

72

 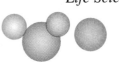

Mask Patterns

Directions: Cut out the mask patterns. Then, cut along the dotted inside line of the masks. Laminate. The clear laminate will simulate a true mask. Punch holes in both sides and add elastic ties.

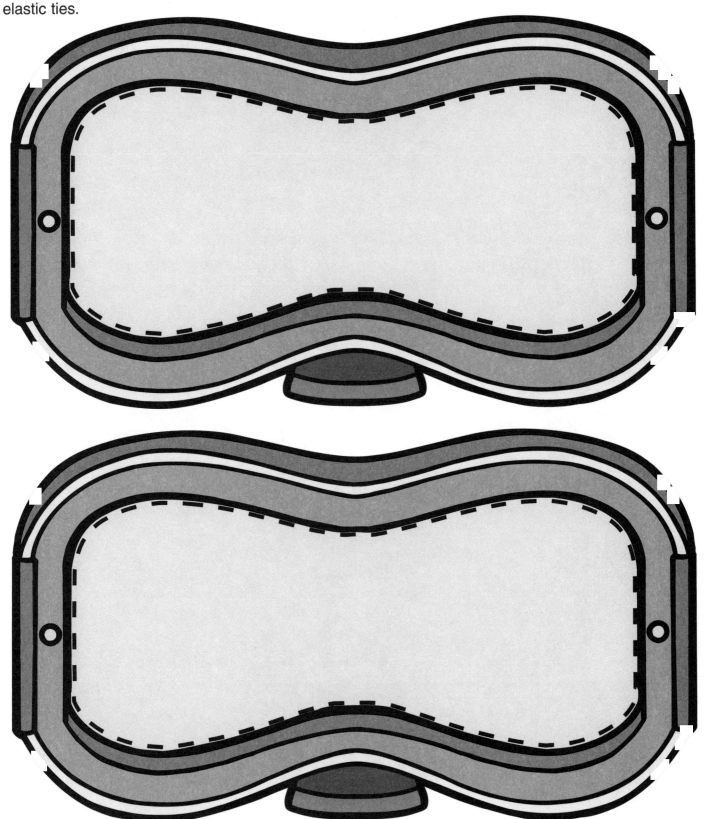

Coral & Fish Cards

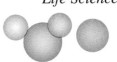

Note to Teacher: Before you cut out the cards for the poster matching activity, you may want to make copies of this page for use in games like "Go Fish" or "Concentration." See suggestions for play in the Lesson section.

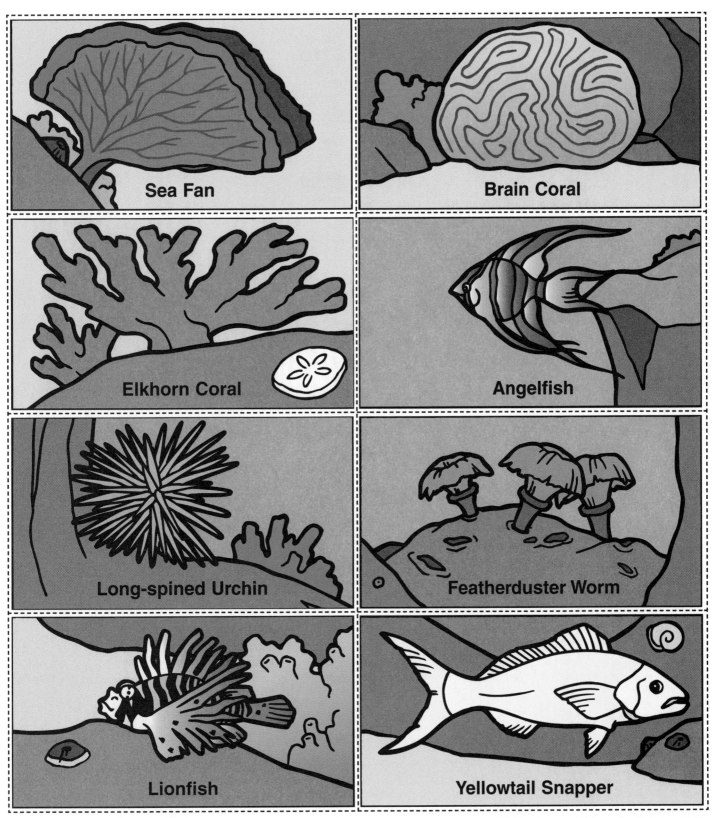

Sea Fan

Brain Coral

Elkhorn Coral

Angelfish

Long-spined Urchin

Featherduster Worm

Lionfish

Yellowtail Snapper

Coral & Fish Cards

Note to Teacher: Before you cut out the cards for the poster matching activity, you may want to make copies of this page for use in games like "Go Fish" or "Concentration." See suggestions for play in the Lesson section.

Clownfish

Parrotfish

Giant Clam

Sea Turtle

Thorny Starfish

Orange Cup Coral

Finger Coral

Sheet Coral

Types of Coral

Fan Coral

Brain Coral

Finger Coral

Sheet Coral

Elkhorn Coral

Orange Cup Coral

Look at the different kinds of coral. Can you draw a picture of your favorite one?

Who Lives in the Reef?

Name: _____

Directions: Look at the pictures below. Cross out the animals that do not live in a coral reef.

 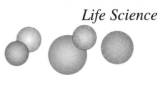
algae	coral
ocean	polyp
reef	scuba dive
skeleton	snorkel

Vocabulary Cards

Coral Reefs   Coral Reefs

Coral Reefs   Coral Reefs

Coral Reefs   Coral Reefs

Coral Reefs   Coral Reefs

Shadows

Children are fascinated by shadows. They love to discover what a shadow is, how it looks, how it changes in the light, what it can do, and how they can make one.

What is a shadow? Light only travels in straight lines. It cannot bend. When a light source is blocked by any object, a dark area is formed behind the object. The dark area is called a *shadow*, or reflected image. A shadow touches the object where it touches the ground.

Can a shadow move? A shadow makes the same movements as the object it reflects. The shadow cannot separate from the object that created it. If the object is moving, or in the air, its shadow follows it.

How big is a shadow? The size of the shadow is determined by two things—the source of light and how close that light source is to the object. The size of the object is also a factor. For example, an adult's shadow would be longer than a child's shadow. And, if an object is placed very close to a light source, it will block out more light. Thus, it makes the shadow larger. If the object is placed farther away from the light source, less light is blocked and a smaller shadow is created.

Shadows created outside on a sunny day are constantly changing lengths. This is because the angle of the rays of the sun changes as the sun travels. This, in turn, causes the shadows to appear to grow or shrink.

How does the time of day affect shadow length? The length of a shadow depends upon the sun's position in the sky. A shadow will be long at sunrise and become progressively smaller at noon, when it is at its shortest. Then, the shadow will become increasingly longer from noon until sunset.

How does the time of year affect shadow length? During winter, the sun is positioned low in the southern sky. It will cast longer shadows at this time of the year than in the summer, when the sun gets progressively higher in the southern sky.

Are shadows clear or fuzzy? A shadow does not show details. Only the outline shape of the object is visible. A shadow may be clear or fuzzy. This depends on how near the light source is to the object, as well as the size of the light source. Children can see examples of this by putting their hand close to a flashlight and seeing the shadow it creates. Then, have them move their hand farther from the flashlight and notice how the shadow changes.

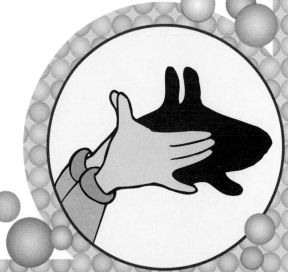

Shadows

Objectives

Students will learn how shadows are created; light does not bend; the sun is a source of light.

Materials

- books about shadows and related materials
- overhead projector
- screen, white sheet, or white butcher paper
- objects to make shadows
- activity card, laminated (page 87)
- copies, "Where's My Shadow?" (page 89)
- backgrounds (pages 91 and 93)
- Shadow Cards, cut and laminated (page 95)
- white construction paper (for background picture)
- sheets of white and black construction paper (approximately 4" x 5"), stapled together at all four corners
- Hand Shadows (pages 97 and 98)
- long sheets of black paper and white chalk
- crayons or markers
- scissors
- glue
- plastic resealable bag for cards

Preparation

Visit the library and collect books related to shadows. When looking for resources, search words might include *light*, *shadow*, and *eclipse*. Related topics might include *Chinese shadow puppets*, *sundials*, *Groundhog Day* (page 90), and *hand shadows*.

If you have parent volunteers, a fun preview activity might include making silhouettes. Place a large sheet of black construction paper on the wall. Seat the child between the wall and the overhead projector. Have the parent trace the child's profile with white chalk or white crayon. If you place these on a bulletin board, students can guess which profile belongs to which student.

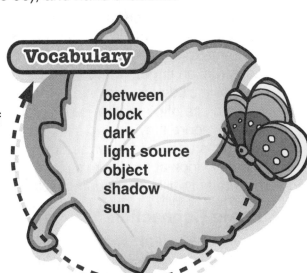

Vocabulary

between
block
dark
light source
object
shadow
sun

Lesson

Give students time to share their knowledge of shadows. Summarize their contributions and explain that they will be building on their knowledge with a variety of activities.

Shadows

Lesson *(cont.)*

Explain to students that a shadow is formed when an object comes between a source of light and the surface on which the shadow will appear. It may be helpful to go outside to demonstrate the concept. Point out that, when standing, a person blocks the sun's light from touching the ground. The person's shape is re-created as a shadow. Allow time for children to practice this concept to internalize it. This might be a good time to introduce a game of shadow tag. One student is designated as "it" and tries to step on another student's shadow. After a little while, have the students change roles and continue playing the game.

Back in the classroom, darken the room. Use an overhead projector as the light source and your hand as the object used to create a shadow. Point out that your hand will be the object that comes between the light and the screen. Choose another object and repeat the process.

Tell the students that they will visit two different centers. (These center activities will work well with pairs of students but can be easily adapted for small groups or individuals.) At the first center, the students will make shadows, using the overhead projector and the collection of objects. If working in pairs, one student can pick an object while the second student tries to guess what the object is by looking at its shadow. The students might also like to duplicate the hand shadows demonstrated on pages 97–98.

At the second center, students will use the backgrounds and shadow cards to practice what they have learned. Be sure to model this activity for them first. Show them where the shadow is in relation to the object pictured on the card, and ask where the sun would be (to the left, to the right, directly above?). Reinforce that the object must be between the sun and the shadow. In pairs, one student can place the sun in position on a background card and the partner could then select the correct object and shadow. They should take turns selecting backgrounds, shadow cards, and placing the sun.

Assessment

Check your students' understanding of shadows by having each student complete "Where's My Shadow?" (page 89). A second assessment activity would be to ask your students to create their own background pages using crayons or markers and white construction paper. Encourage them to be creative in setting their scenes and to be sure to include the sun. Next, they will draw and color the figure of a boy or girl on a white piece of paper that has been stapled to a black piece of paper. When they cut out the figure they have drawn, they will be simultaneously cutting out its shadow. Instruct students to glue the drawn figure onto the background. They should then carefully look at the figure's placement in relation to the sun, and position the shadow accordingly. Have an adult check before the shadow is glued to the paper.

Home Connection

Encourage parents to take out flashlights and have their children demonstrate how to make shadows. You might like to send home a copy of the hand shadows on pages 97–98 for the family to practice together.

Shadows

Station 1

Place objects between the light source and the screen to create shadows.

Station 2

1. Take a background, a shadow card, and the sun card.

2. Place the shadow and sun cards in the correct position on the background.

3. Check to see if the object in the picture is between the sun and the shadow.

Where's My Shadow?

Directions: Complete the picture. Draw shadows for the boy and the girl.

A Shadow Holiday

Name:_____

Groundhog Day is a popular tradition that is observed every February 2. It is a day when the groundhog emerges from his hole after a long winter slumber. Legend has it that if he sees his shadow, he will go back into his hole, indicating six more weeks of winter. If he does not see his shadow, he will exit his hole and celebrate this sign of spring! Have you ever seen a groundhog?

Shadows

Background 1

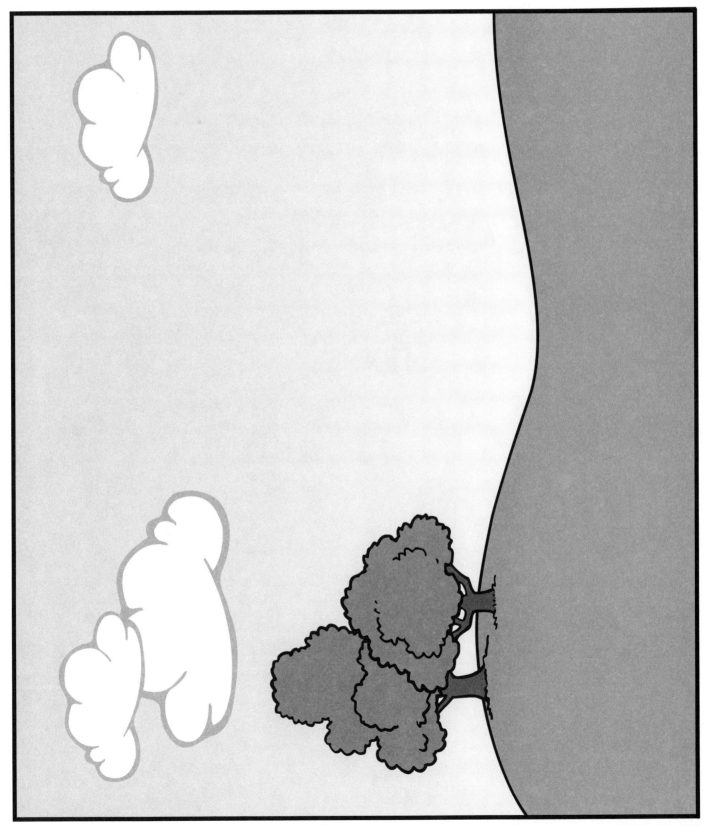

Shadows

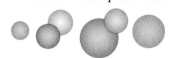

Background 2

Shadows Cards

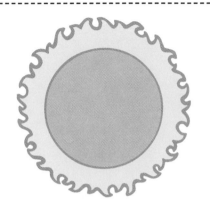

Hand Shadows

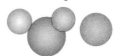

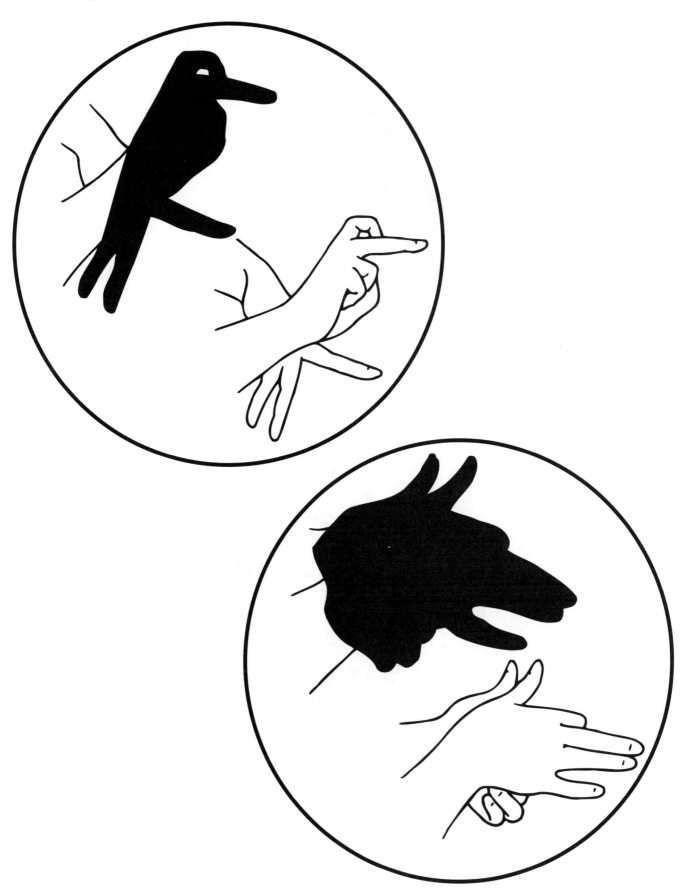

Hand Shadows

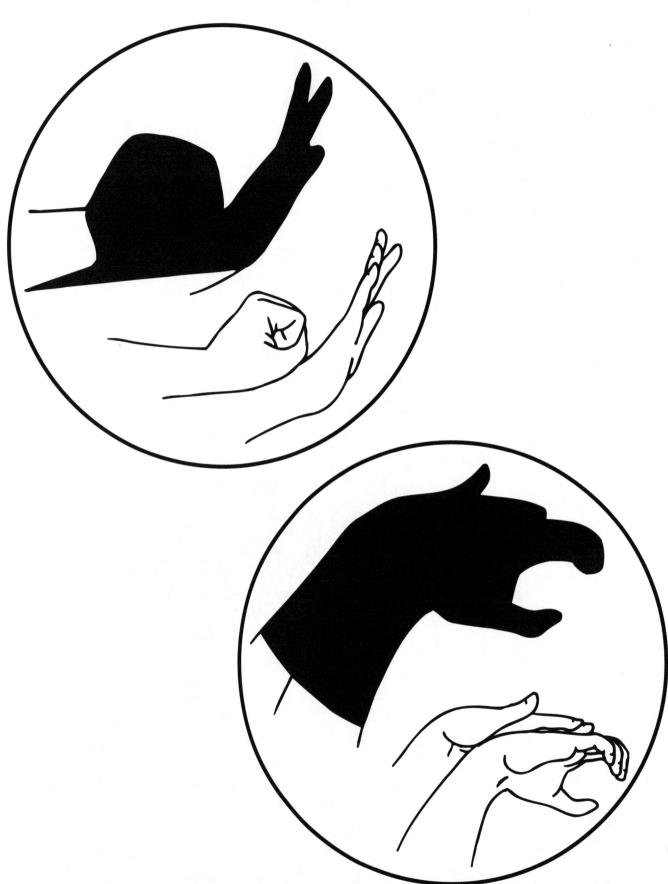

 # Vocabulary Cards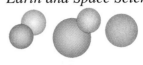

block	dark
object	shadow
sun	between
light	source

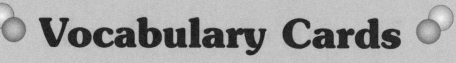 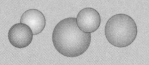
Shadows

Shadows

Shadows

Shadows

Shadows

Shadows

Shadows

Shadows

Lakes and Rivers

What is a lake? A *lake* is a landlocked body of still water that sits in a hollow in the ground. Lake water comes from glacial melts or rainfall and can be fresh or salty. Large lakes are often called *seas*. Small lakes can be called *ponds*.

What is a river? A *river* is fresh, moving water fed by other rivers, melting glaciers, and rainfall. As it flows, it carves out a path in the ground. Small rivers are called *streams*.

What do lakes and rivers look like? Lakes and rivers can be different sizes, shapes, lengths, and depths due to weather conditions, erosion, and land and glacial movement.

What is the water cycle? The heat of the sun and the pull of gravity keep Earth's water in circulation. Water evaporates and rises in the form of water vapor. The water vapor cools, collects into clouds, condenses, and returns to the ground as moisture in the form of rain or snow. Then, the cycle begins anew. The same water has been in a continuous cycle since the time of the dinosaurs!

What is erosion? Rivers are constantly shaping the surface of the earth through a process called *erosion*—the process of "wearing away" a surface. Flowing water picks up particles of rock and soil, plants, insects, and animal life, and then deposits the material or debris downstream.

Are rivers and lakes viable habitats? A wide variety of plants and animals live in both lakes and rivers. Salmon and trout thrive in the fast-flowing current of the upper river. As the river slows down, toward its middle and especially at its end, it deposits nutrient-rich sediment ideal for plant growth. Microscopic algae support a wide variety of insects. Reeds, rushes, and other tall plants provide cover and nesting places for a wide variety of birds, fish, and small mammals, including otters, turtles, and ducks. Large mammals, such as the hippopotamus, also live on rivers.

A lake is an ecosystem in which all the plants and animals depend on one another for survival. Lakeside land is usually marshy. Tall reeds and grasses provide shelter for ducks, like the wood duck, and small reptiles, such as frogs. Catfish and pike are just two of the many varieties of fish found in lakes, and birds like the kingfisher eat insects that live there.

What is a waterfall? A *waterfall* is a vertical drop over which a river falls. Niagara Falls tumbles over the edge at a rate of 50 miles per hour!

When is a lake salty? When lakes have no outlet, minerals, like salt, will build up. In hot, dry climates, water evaporation between rainfalls can leave a lake salty.

River and lake records: The world's longest river is the Nile, at 4,145 miles. The Caspian Sea, in Central Asia, is the world's largest natural lake, covering an area of 143,600 square miles.

Lakes and Rivers

Objective

Students will be able to identify the differences between a river and a lake, and will be introduced to the water cycle.

Materials

- books, pictures, maps, posters
- water source
- sand table center
- sand or dirt
- funnels, hose, pitchers
- rocks, pebbles, leaves, and gravel
- plastic animals, trees, and vehicles
- Water Cycle poster (page 107), laminated
- copies, "The Water Cycle" mini books (pages 109–112)
- "Lakes and Rivers" worksheet (page 108)
- activity card, laminated (page 105)

Preparation

Arrange nonfiction books about lakes and rivers in a center area. Display maps, posters, and assorted pictures that include lakes and rivers. Share some of these items with students to introduce the concept of lakes and rivers.

Lesson

Begin by explaining that there is water all around us. Explain that water follows a specific cycle. Share the poster on page 107. Discuss the cycle. Allow time for students to work on and share their individual water cycle mini books.

Brainstorm a list of inland water forms—rivers, lakes, streams, ponds, lakes, puddles, waterfalls, hot springs, etc. Add to the list as more information is learned. Ask your students if they have ever visited a river or a lake. What kinds of things did they see? What did they do?

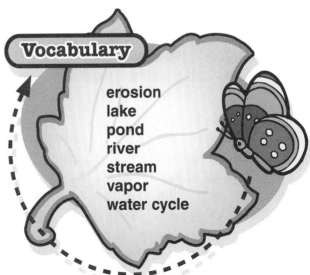

Vocabulary

erosion
lake
pond
river
stream
vapor
water cycle

Share a large map of the United States or of the world. Point out the oceans. Then, ask students if they can identify other water areas. Guide them to note similar blue areas on the map that represent lakes and rivers. Or, show students a topographical map depicting specific lakes and rivers in your area. Trace the paths of rivers and the shapes of lakes. Note that lakes are more circular and rivers are more linear. (You can drive around a lake, but not a river.)

Lakes and Rivers

Lesson *(cont.)*

Read from two or three nonfiction books (pictures are a necessity) about lakes and rivers. Discuss the facts included on the teacher Facts page, specifically what happens to rocks and pebbles as they travel down a flowing river (*erosion*).

Tell your students that they will be making rivers and lakes in sand (or soil). Model for them how to work at the center. (Weather permitting, this center would be easier in an outdoor area.) Show them how to make one end of the sand area higher to form a mountainous area. Mention that they can create larger crevices for rivers and smaller trenches for streams. Using a hose, pitchers of water, and funnels, have students begin a river source at the top and then watch the water move downward. Demonstrate how to pour carefully. Shallow holes will make ponds and wider holes will make lakes. Incorporate rocks, pebbles, leaves, etc. If possible, slightly elevate one end of the sand table. Create an ocean at the bottom to show how the water eventually ends up in the ocean. Students can add plastic animals and trees.

Assessment

Make duplicates of the mini book pages to use as cards in a center. Have students arrange the cards in the correct order to demonstrate the water cycle.

Observe the "lakes and rivers" created in the sand area. Are the distinctions between the lakes and the rivers obvious? Can the student point out where he/she made a lake or river? Add small plastic cars or animals and have students indicate how the animal or car can travel around the lakes and alongside the rivers.

Have the students complete the "Lakes and Rivers" worksheet (page 108). Students will begin at the top of the mountain and illustrate a river's path and draw a lake.

Home Connection

Here is a good way to demonstrate erosion with your students. Send these instructions home for families to do together or conduct the activity in your classroom.

Making Sand

Materials
- glass of water
- rocks and pebbles
- coffee can (or similar) with lid
- tape

Directions

Fill a clean can with clean water halfway full. Then, add four or five small rocks. Put the lid on the container and tape it shut. Shake the container for five or ten minutes. Do this every day for at least a week. The more time spent shaking, the more erosion (the wearing away of the rock) there will be.

When the shaking period is over, open the can and pour the water into a clear glass or cup. Is the water still clear or has it gotten dirty? Notice if the pebbles have changed shape or size. Check the bottom of the can; do students feel or see any rough sand or grit? Review that this is similar to the current in a river; the water makes rocks hit one another, and sand is formed.

Lakes and Rivers

1. Arrange the sand to create a mountain at one end.

2. Form channels for rivers and holes for lakes.

3. Pour water into the lakes and rivers.

4. Watch water run down the river and sit in the lake.

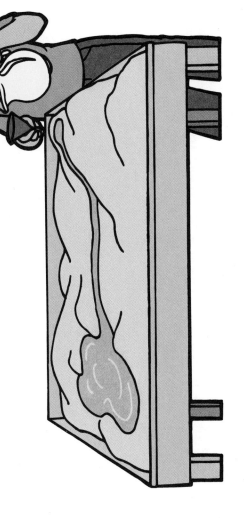

The Water Cycle

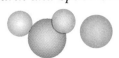

1. The sun heats the water. Water turns into vapor. This is called *evaporation*.

2. Vapor cools as it rises and forms clouds. This is called *condensation*.

3. Rain falls from heavy clouds. This is called *precipitation*.

4. Water collects on the ground, in rivers, lakes, and oceans.

Lakes and Rivers

Name:_____

1. Illustrate a river's path.
2. Add a lake.
3. Color the water blue.

4. Color the land green and brown
5. Add trees, animals, and vehicles.

 # Mini Book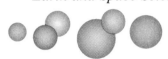

Directions: Cut the mini book pages along the dotted lines. Arrange the pages in order. Staple twice on the left side. Illustrate your book.

The Water Cycle

By

_____

 # Mini Book

Water is heated by the sun.

1

Evaporation

The water turns into steam, or *vapor*, and goes up into the air.

2

Mini Book

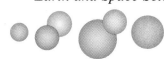

Condensation

The vapor turns into clouds.

3

Precipitation

Heavy clouds make rain.

4

Mini Book

Rain collects in rivers, lakes, and oceans.

5

The sun comes out and the cycle begins again.

6

 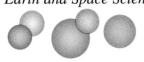
erosion	lake
pond	river
stream	vapor

water cycle

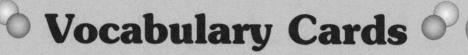 

Vocabulary Cards

Lakes & Rivers   Lakes & Rivers

Lakes & Rivers   Lakes & Rivers

Lakes & Rivers   Lakes & Rivers

Lakes & Rivers

Stars

What is the sun? The sun is the largest known object in our solar system. It is a star—a huge ball of burning gas. It is so hot that it can provide light and heat for our planet from 93 million miles away! The sun seems huge to us only because it is the closest star to Earth. The sun is absolutely essential for life on Earth. It takes one year for Earth to travel around the sun.

What is a star? Stars are hot balls of glowing gas that shine for millions of years. Every 18 days, another star forms in our galaxy. Scientists think that there are more stars in the sky than there are grains of sand at the beach! Groups of millions of stars are called a *galaxy*. We live in the Milky Way Galaxy.

Most stars are so far away that they look like tiny dots of light in the sky. We cannot see them in the daytime because the light from our sun is brighter and closer than the light of the other stars.

What is a constellation? Stars are fixed in the sky; they do not move. They form consistent, recognizable patterns. These patterns have been given names over the years by farmers, poets, and astronomers. Each pattern is called a *constellation*. Since stars do not move, different constellations are observed in different parts of the world. For instance, in the United States, located in the Northern Hemisphere, you may have seen the Big Dipper or Orion. In Australia, located in the Southern Hemisphere, the Southern Cross and the Triangle are visible. It takes practice to find them. Years ago, before modern technology, sailors used the position of the stars to help guide them in their travels.

What is rotation? Earth spins completely around every 24 hours. This movement is called *rotation*. Earth is always spinning, although we cannot feel it. It is brighter and warmer on that part of Earth that is facing the sun and receiving its light. This is daytime. At night, it becomes colder and darker when that same part of the earth rotates, or turns away, from the sun. When it is daytime in half of the world, it is nighttime in the other half.

As Earth rotates, it also revolves around the sun. It takes one year to make a complete trip around the sun.

Why does the amount of daylight change during the year? Earth travels around the sun on a set path, called an *orbit*. Its position on its axis and its location on its trip around the sun will determine the length of daylight. (The axis is an imaginary line connecting the North and South Poles.) During the summer, there are more hours of daylight. In wintertime, there are more hours of darkness.

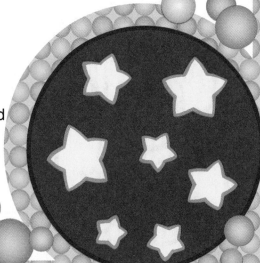

Daytime/Nighttime

Name: _____

Directions: Look at each picture. Decide if the sun would be shining or the moon and stars would be glowing. Circle your choice in each box.

Stars

Objectives

Students will recognize the differences between night and day and how they occur (rotation); they will learn the basic characteristics of our sun and other stars.

Materials

- books and pictures about the sun, stars, day and night (Include page 121.)
- constellation maps
- blankets or sheets
- large table(s)
- flashlights
- templates, cut and laminated (pages 123, 125)
- "Rotation and Revolution" mini poster (page 120)
- black construction paper
- Vocabulary Cards, cut and laminated (page 127)
- box cutter
- activity card, laminated (page 119)
- copies, "Daytime/Nighttime" (page 116)
- copies, "Day and Night" (page 122)
- scissors, glue
- two different sized balls
- globe (optional)
- "Constellations" mini poster (page 125)
- Optional: blank calendar pages

Preparation

Arrange nonfiction books about day and night, the sun, and other stars in a center area. Display constellation maps, and pictures of the sun, moon, and stars.

Place a large banquet-size table in an open space. Place blankets over the table to make the center dark.

Lesson

Ask your students what they already know about the sun and the stars. How does the sun relate to daytime and nighttime? Then, determine what other things the class wants to learn about the topic.

Discuss the differences between day and night. Take your students outside and observe what it is like during the day. Then, ask them to imagine that they are in the exact same spot at night. What would they see? How would it be different?

Make a Venn diagram listing activities that the students regularly do during the day, at night, or at both times. As a class, complete "Daytime/Nighttime."

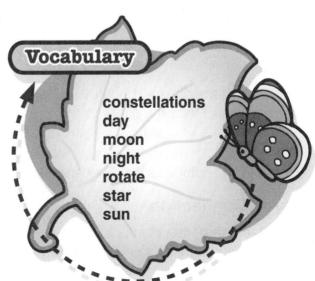

Vocabulary

constellations
day
moon
night
rotate
star
sun

Stars

Lesson (cont.)

Create a model of Earth and the sun (a tennis ball and a basketball). Mark a spot on the smaller ball to suggest the school's location on "Earth." Have one student hold the larger ball to represent the sun. Another student will hold the smaller ball to represent Earth. (See page 120.) Explain that each day, Earth rotates (spins around) once. During this rotation, there are times when "the school" is facing the sun and receives its light. This is daytime. When it is nighttime, "the school" will be in the dark because it does not face the sun and cannot receive its light. Have "Earth" slowly rotate. When the children understand the concept of rotation, have the "rotating Earth" slowly make a trip around "the sun." Remind the students that it takes Earth 24 hours, or one day, to make a full rotation. It takes one year for Earth to make a complete trip around the sun (a *revolution*). Remind "Earth" to continue spinning as he/she revolves around the sun. Remind "the sun" to stay still.

Discuss stars. Mention that the sun is the closest star to us, so its light shines brighter than the more distant stars. Therefore, we do not see the other stars during the day even though they are there in the sky.

Remind students that stars do not move. Share illustrations of constellations. (See page 125.)

Clarify for your students that the moon is a ball of rock, not a star. It glows because it reflects the light from the sun. The different shapes (phases) of the moon are created when different parts of the moon reflect the sun's light.

Explain that they will be using flashlights in a darkened area. (Make certain that students who may be uncomfortable with this are helped or excused.) Show students the flashlights and demonstrate how to turn them on and off. Hold up some of the templates and explain that students will take turns holding different templates over the flashlight while under the table. Darken the classroom and demonstrate how to do this. Show them how the size of the reflected image changes if the template is held close to the light or farther away. Show them how the size of the reflected image also changes depending on the flashlight's distance from the reflective surface. If the flashlight is closer to the table, the shape becomes smaller, etc.

Hints

If your students have a difficult time holding the filters and the flashlights, round up more flashlights and tape the filters to the flashlights.

Assessment

Students will complete the "Day and Night" worksheet on page 122.

Home Connection

Give each student a blank calendar page. Encourage them to look at the moon each night and draw a picture of it on the appropriate day. At the end of the pre-determined time (one week, one month. . .) have the children bring in their observation sheets to share with the class.

Stars

1. Take a flashlight and a template with you under the table.

2. Turn on the flashlight. Place the template over the flashlight.

3. Shine the flashlight on the underside of the table.

4. Can you identify the shape or picture?

Rotation and Revolution

Our Solar System

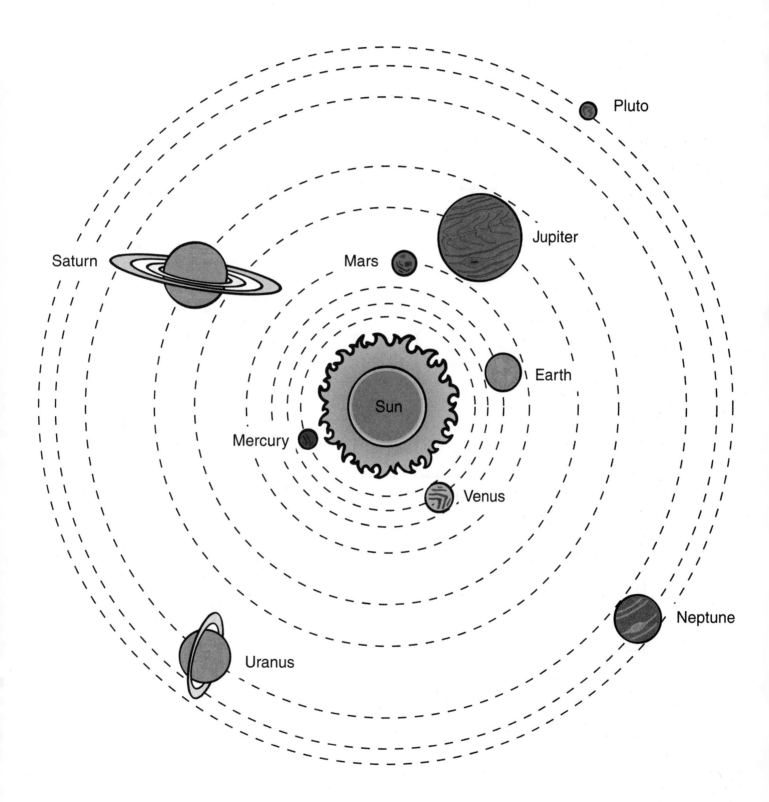

Pluto

Jupiter

Saturn

Mars

Sun

Mercury

Earth

Venus

Neptune

Uranus

Day and Night

Name: _____

Directions: Cut out the words below. Paste them in the box under the correct picture.

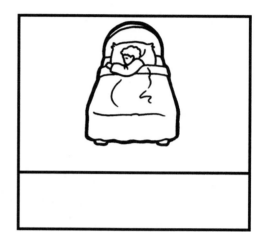

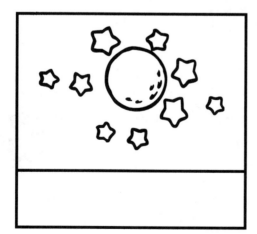

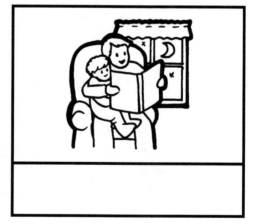

night	day	night
night	day	day

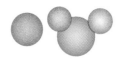

Flashlight Shapes

Teacher Directions: Place the shapes below on black construction paper. Use box cutters to cut out the designs. Laminate. (**Safety Note:** Students should not use box cutters.)

1.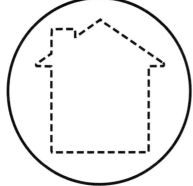

2.

3.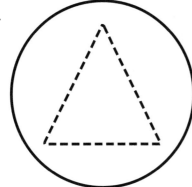

4.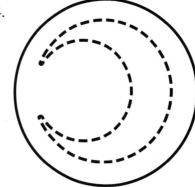

5.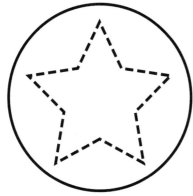

6.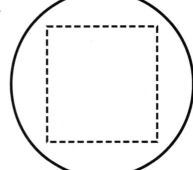

7.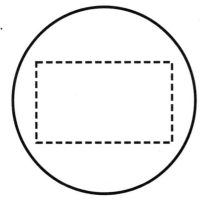

8.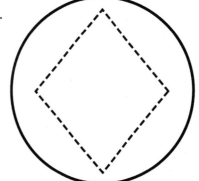

Constellations

Northern Hemisphere

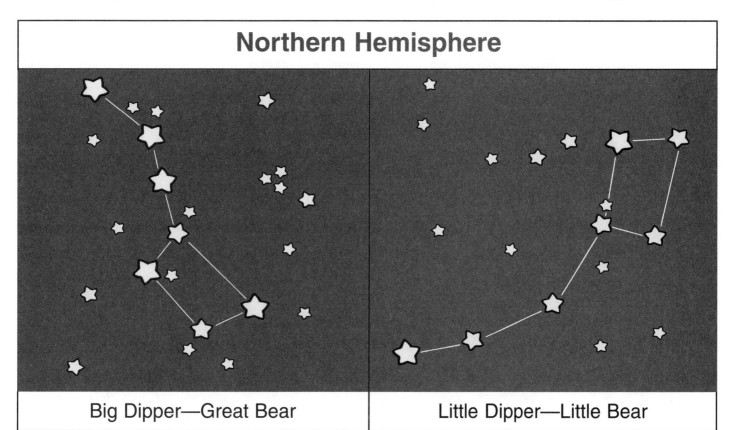

Big Dipper—Great Bear

Little Dipper—Little Bear

Southern Hemisphere

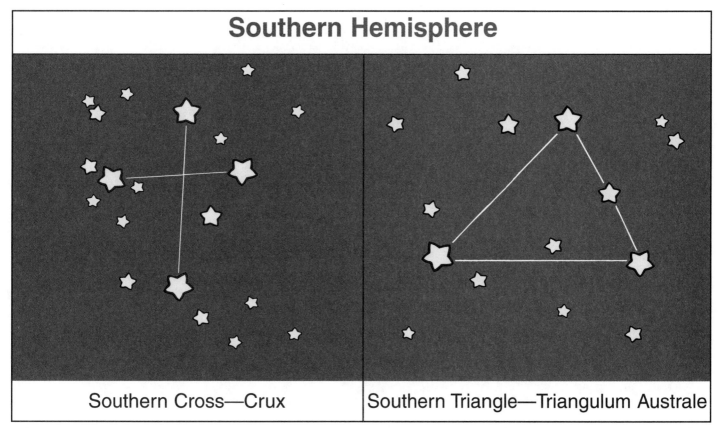

Southern Cross—Crux

Southern Triangle—Triangulum Australe

 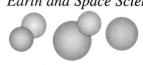
sun	star
moon	rotate
day	night

constellations

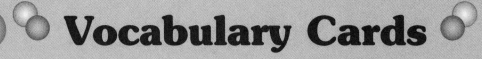

Vocabulary Cards

Stars

Stars

Stars

Stars

Stars

Stars

Stars

Trees and Wood

What is a tree? Trees are the largest of all plants. They are alive. All trees begin as a seed and must have food, water, and light to grow. Trees have roots, bark, a trunk, branches, and leaves or needles. Trees can live longer than human beings or animals.

What purpose do the roots serve? Roots anchor the tree, supporting it in the ground so that it can stand upright. They also absorb the water and minerals from the soil that the tree needs to survive. Roots also hold soil in place which helps prevent erosion and flooding.

What is the job of the tree's trunk? The trunk supports the tree and connects the roots to the tree's branches, transporting the food taken in by the roots. The branches, in turn, deliver food to the leaves.

How does the bark help the tree? The outer bark helps the tree retain moisture when it is hot and dry outside. Bark insulates the tree from the hot and cold weather and helps protect the tree from insects. Nourishment, collected by the roots, travels throughout the tree via the inner bark, or *phloem*.

What is photosynthesis? *Photosynthesis* is a process by which the leaves on a tree use energy from the sun to convert carbon dioxide in the air and water from the soil into sugars that feed the tree. During this process, the tree releases oxygen into the air. This is very important since all humans and animals need oxygen to survive.

Are all trees the same? No! Between 60,000–70,000 different species of trees have been identified. Some trees provide fruits or nuts. Some trees flower. Some are green all year round *(evergreen)*, while others drop their leaves each winter *(deciduous)*.

Why are trees important? Trees produce oxygen, necessary for survival. They provide food for humans and animals (nuts, fruits, seeds). They offer shade and shelter for a variety of animals (owls, raccoons, birds, squirrels). They also provide firewood for heating.

What is wood? Wood is a layer of tissue, found beneath a tree's bark, which is cut and dried for a multitude of uses. Just as there are many different kinds of trees, there are many different kinds of wood. There are differences in color, texture, and density.

Wood is used to produce a wide range of products including furniture, lumber for building homes, toys, fences, paper products (newspapers, boxes, paper bags), mulch, and even crutches!

Where is the biggest tree? A giant sequoia, called the General Sherman, is located in the Sequoia National Park in California. It is over 275 feet tall, measures over 82 feet in circumference, and weighs about 6,700 tons!

How old is the oldest tree? The oldest living thing on Earth is a 4,700 Bristlecone pine tree in California.

A Tree Celebration

A holiday to celebrate trees? Arbor Day, celebrated since 1872, got its start when the state of Nebraska urged its settlers to plant trees on the largely treeless prairies there. They knew that trees would provide shade, fruit, and fuel, as well as beauty. Find out when your state celebrates Arbor Day by checking ***www.arbor-day.net***. You can enjoy the day by planting a tree, caring for an existing tree, or just appreciating the beauty of trees around you.

Directions: Color the pictures. Cut them out and arrange them in order.

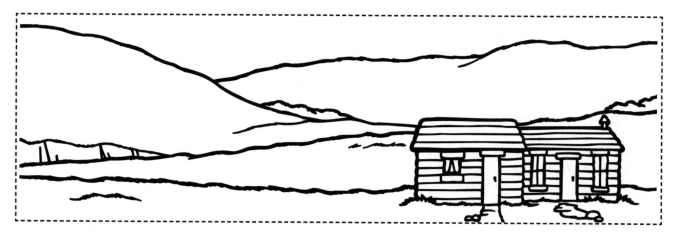

Trees and Wood

Objective

Students will learn the basic parts of a tree and come to appreciate the wide variety and versatility of this precious resource.

Materials

- pictures, books of trees
- wood samples
- tree products and items made from wood
- peeled crayons
- masking tape
- drawing paper
- copies, "Cross Section of a Tree" (p 139)
- Vocabulary Cards
- copies, "Parts of a Tree" (p 140)

Note: * See each individual center for a list of specific materials for each activity

Preparation

A few weeks prior to establishing this center, visit several local lumberyards, tree trimmers, and contractors to collect a variety of wood samples. Write the name of each type of wood on the sample with a piece of masking tape. Bring a container with a lid to collect scrap sawdust.

At the science center, display the wood samples and pictures of different kinds of trees (pages 141 and 143), books, and posters about trees and wood. Arrange tree products (fruit, nuts, olives) and items made from wood (frames, pencils, paper goods, etc.) in the center.

Lesson

Go outside and have children observe and touch trees. Discuss the different parts of a tree. Point out the leaves, branches, trunk, roots, and flowers or fruit, if appropriate. Discuss similarities and differences in the trees and wood you have observed. Have each student select a tree and securely tape drawing paper to the bark. They will rub a crayon back and forth across the paper, using the flat, peeled side. What do they observe about the pattern? Is it different from other tree rubbings that their friends made? Do they know what kind of tree it is? Display the bark rubbings in the classroom.

Back in the classroom review the parts of the tree with the worksheet, "Parts of a Tree," (page 140). Explain that you are now going to focus on a certain part of the tree—the part under the bark. Share the information about tree rings on page 139. Show different wood samples and pictures of the types of trees from which they came. Brainstorm different uses for wood including shelter, furniture, fences, containers, etc. What items in the classroom are made from wood? Record student ideas on a large piece of chart paper.

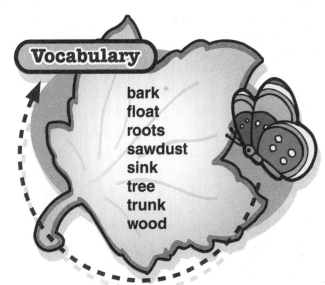

Vocabulary

bark
float
roots
sawdust
sink
tree
trunk
wood

Select any or all of the centers listed for your science center. One center will be working with the scraps of wood to see how they react to water; another center will be observing how sand paper changes wood. At the last center, students can make a design using sawdust. Assessment pages are included for each center.

Trees and Wood

Center One – Sink and Float

Materials: large bucket of water; variety of wood samples; objects to be used for weight (paperclips, metal washers); copies, "Sink or Float" worksheet (page 134); rubberbands; activity card (page 133).

Directions: Students will choose a piece of wood, place it in the water, and observe what happens. Encourage the students to think about the following: *Does it float? Does it stick to another piece of wood? Does it change color when it gets wet? Will it sink if it stays in water for a long time and becomes waterlogged? What happens when you add a weight* (attach paperclips, washers, etc, to the wood with rubberbands)?

Assessment: Students will record their observations on "Sink or Float" worksheet.

Center Two – Smooth and Rough

Materials: variety of wood samples; three grades of sandpaper: fine, medium, rough; copies, "Rough and Smooth" worksheet (page 136); activity card (page 135).

Directions: Number the sandpapers from 1 (rough) to 2 (medium) to 3 (fine). Use the roughest paper first. Demonstrate how to use the sandpaper, gently rubbing back and forth with the grain of the wood. Rubbing the edges works well for observing change. Before they begin, have them feel the wood texture. Then, tell them to rub and count to 20. Have them feel it again with their fingers. *Does it feel different? Is it smooth or rough?* Continue this same process until you get to the final sandpaper. Tell your students to notice the difference between the way in which the wood felt when they first started and the way it feels now. It should be very smooth. Notice any changes in the shape. Point out that the wood that came off is now in the form of very minute pieces (*sawdust*). Share with your students that woodworkers use sandpaper to shape and finish fine wood pieces. Add the sanded wood pieces to the block area or make collages.

Assessment: Students will complete the "Rough and Smooth" worksheet.

Center Three – Sawdust Art

Materials: sawdust; glue; paintbrushes; small containers; heavy, black construction paper; activity card (page 137).

Directions: Slightly water down the glue and place in open containers with brushes. Have students paint a picture with the glue and then sprinkle sawdust over the picture. Tell them to count to 20 and then carefully remove the extra sawdust. Be sure to model how to gently turn their paper to the side to then carefully pour the extra sawdust back into the container.

Home Connection: Students will make a list of at least 10 things in their home that are made of wood. They might also like to try the bark rubbings with trees in their neighborhood. Can they find trees that are different from the ones surrounding the school? Use the worksheets included in this unit to reinforce concepts being taught.

Trees and Wood

Sink or Float

1. Choose a piece of wood.

2. Place it in the water.

3. Observe what happens.

4. Use the recording page to draw or write about your observations.

Sink or Float

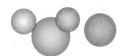

Name: _____

Directions: Draw a picture of what happened to the wood when you put it in the water. Did it sink or float? Circle the correct answer at the bottom of the box.

Sink   ## Float

Directions: Draw a picture of what happened to the wood when you added weight. Did it sink or float? Circle the correct answer at the bottom of the box.

Sink   ## Float

Trees and Wood

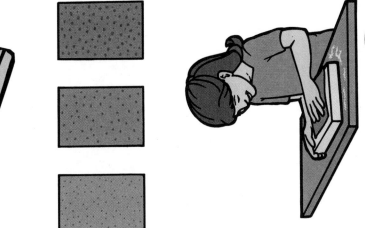

Rough and Smooth

1. Choose a piece of wood.

2. Rub sandpaper #1 on the wood along the grain. Count to 20. How does it feel?

3. Rub sandpaper #2 on the wood along the grain. Count to 20. How does it feel?

4. Rub sandpaper #3 on the wood along the grain. Count to 20. How does it feel?

 # Rough and Smooth

Name:_____

Directions: Color the items that are smooth. Cross out the items that are rough.

Trees and Wood

Sawdust Art

1. Paint a picture with the glue.

2. Sprinkle sawdust over the picture.

3. Slowly count to 20.

4. Carefully shake the excess sawdust back into the container.

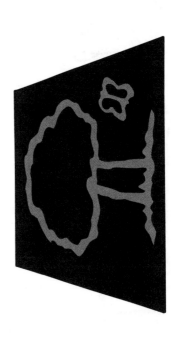

Wood Products

Directions: Look at the pictures in each row. Cross out the item that does not belong. Color the pictures of wood products.

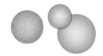 # Cross-Section of a Tree

Name:_____

Each ring in this cross-section represents one year in the life of the tree.

If you were a tree, how many rings would you have?

I am _____ years old.

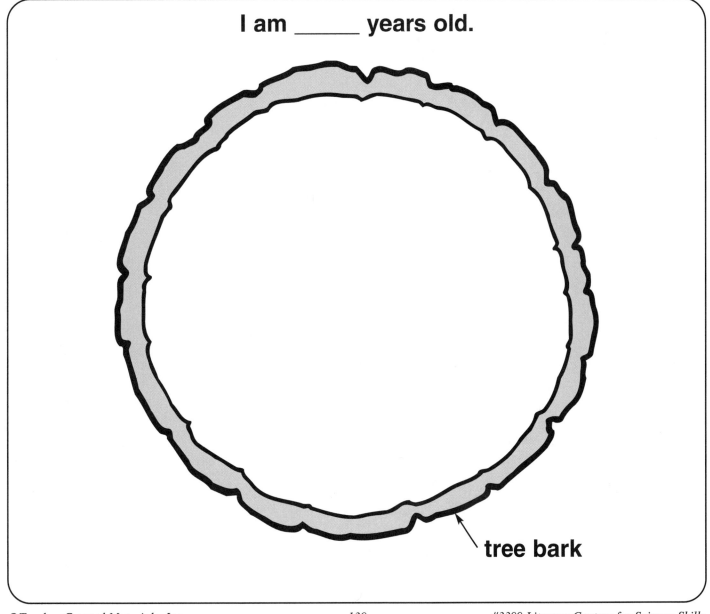

tree bark

 # Parts of a Tree

Name: _____

Directions: Cut out the labels and use them to label the tree.

| leaves | branches | roots | trunk |

Trees and Leaves

Maple

Oak

Palm

Cedar

Physical Science

Trees and Leaves

Weeping Willow

Spruce

Orange

Walnut

 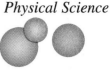
bark	float
roots	sawdust
sink	tree
trunk	wood

Vocabulary Cards

Trees & Wood	Trees & Wood
Trees & Wood	Trees & Wood
Trees & Wood	Trees & Wood
Trees & Wood	Trees & Wood

Water Around Me

What is water? Water is a clear, colorless, odorless, and tasteless liquid. It is composed of two elements—hydrogen and oxygen. (2 parts hydrogen + 1 part oxygen = H_2O) Water is all around us. It covers 75% of Earth's surface. Water can be a solid (ice), a liquid, or a gas (vapor). It is the most common substance on Earth. Water comes to Earth in the form of rain, snow, hail, or other forms of precipitation.

Where is water found? Water is in the air, in the ground, in rivers, lakes, and oceans. Much of the water on Earth's surface is in the oceans, which are salt water. When ocean water evaporates, the salt stays behind. Most rivers and lakes are filled with fresh water.

Water can be found in living things as well. The human body is approximately two-thirds (65%) water. An elephant is 70% water. Fruits and vegetables also have high water contents (ear of corn–70%; potatoes–80%; tomato–95%).

Why is water important? Water is essential for most plant and animal survival. In humans, water regulates body temperature, lubricates joints and tissues, transports nutrients throughout the body, and prevents dehydration. While a person can survive for a whole month without food, he/she could go only 2–5 days without water. In addition, water provides essential habitats for a large variety of fish, reptiles, birds, and mammals.

How is water used? In addition to our need of water for basic survival, it has an endless number of uses. In everyday life, we use water to wash things and to cook food. Firemen use water to douse fires. Water serves as a transportation pathway for a wide range of watercraft. Water is a great source of recreation from swimming, fishing, and boating to skating and skiing. Water is used to irrigate fields that provide our food, and for small gardens, fields, and golf courses that provide great pleasure.

What happens when water freezes? The molecular structure of water actually causes it to expand when frozen. It is not uncommon in very cold places to hear that water pipes have broken. The frozen water inside the pipes expands and eventually breaks the pipes! Frozen water is lighter than water, which is why ice floats in water.

Is water dangerous? Water can be very dangerous. People can drown in very small amounts of water. Floodwaters create serious damage to homes, highways, and crops. Massive tidal waves or tsunamis can destroy entire towns or villages. Polluted waters can cause serious illnesses and destroy vital habitats. Boiling water can cause serious burns, and running water can erode necessary shoreline.

Water Around Me

Objectives

Students will understand that water is all around us; we need it to survive in our everyday lives; it has a variety of uses.

Materials

- books, posters, pictures of water, water sports, water habitats, water uses
- soap
- water
- different-sized ice cubes and ice blocks
- bowls
- paper, pencils
- baby food jars
- dirt, sand, flour, salt, and oil
- copies of appropriate worksheets, pages 152, 154, 156, 157, and 158.
- soap, bath toys, cans of soup, envelopes of *gelatin*, packets of seeds, tea bags
- Vocabulary Cards, cut and laminated (page 159)

Preparation

Tell your students that you are organizing two new science centers. Place the following items in a box or on a table and challenge them to guess what all the items have in common: soap, bath toys, cans of soup, envelopes of *gelatin*, packets of seeds, tea bags, etc. Once the students have concluded that each item needs water for use, encourage them to find additional items to add to the collection. Gather nonfiction books about water, water habitats, waterways, water uses, and water sports to place in the center.

Brainstorm with your students on how water is used in their everyday lives. Talk about how it is used to brush their teeth and to bathe. Water is needed to grow fruits and vegetables. Point out how water is used when preparing foods, for cleaning the dishes afterward, etc. Ask your students to come up with other ideas and list them on the board or on chart paper.

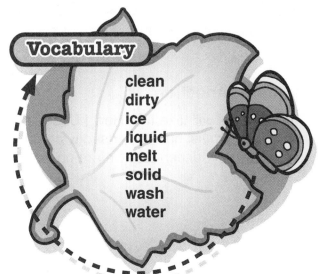

Vocabulary

clean
dirty
ice
liquid
melt
solid
wash
water

Talk about the importance of hand washing. Model how to thoroughly wash and dry your hands. A handy guide is to tell your students that, after they wet their hands, they should continually rub their hands with soap for as long as it takes for them to sing "Happy Birthday." When they have completed the verse, it's time to rinse and dry. Color and review the worksheet on page 154.

Water Around Me

Lesson

Before introducing the centers to your students, spend a little time with water. Talk about how things like dirt, salt, and flour change when water is added to them. Can they think of things they add water to and in what ways they change? (You may want to refer to the collection of items introduced in the Preparation section.)

In baby food jars, have small amounts of dirt, sand, flour, salt, and oil. Add a small amount of water to the first jar of dirt, and discuss what happens to the dirt. Be sure all your students can see the results. Do the water and the dirt mix? Is the water still clear? As a class, complete the "Mixing with Water" worksheet (page 157) as each mixture is used. Continue this same procedure with the sand, flour, salt, and oil. Set the jars aside. Revisit them after an hour or so and discuss how the different substances settled in the water.

Tell your students that water is not always a liquid. It can also be a gas or a solid. Explain that when water is frozen, it becomes a solid and is called *ice*. Ask your students to think about situations in which frozen water is used. Review the worksheet on page 156 with your class.

Show your students two blocks of ice—one ice cube and one larger block frozen in a pint container. Check a clock and write down the time the ice was removed from a freezer. Place each ice sample in a separate bowl and set them aside in your room.

Tell students that they are going to make two estimates. First, they will estimate how long it will take for the ice to become a liquid. Second, they will estimate how much water the ice cube will make when it becomes a liquid (melts).

Have the students write their predictions down on a sheet of paper. Show them a measuring cup to help them with predicting the water amount. Check the bowl every half hour and record the results on the board. Which melted more quickly, the small ice cube or the larger one? Which produced more liquid? What conclusions can be drawn?

Assessment

There are a number of assessments sheets included for "Water Around Me." Some will enhance discussions about different aspects of water. Others will reinforce specific activities.

Home Connection

The students will take the "How I Used Water" recording sheet home (page 158). An adult can help them list how they used water in just one evening. As a follow-up, you may want to discuss how dependent we are on water and ways in which we can conserve it.

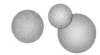

Water Around Me

Center One – Washing Center

Objective: The student will understand the importance of keeping an object clean.

Materials

- large tub of water, or small wading pool for washing
- small tub of water for rinsing
- 1–2 tablespoons of dishwashing detergent
- sponges or rags
- towels
- items to wash: toys, play dishes, etc.
- laundry basket for storing items
- activity card, laminated (page 153)

Directions: Discuss with your students how water is used to wash and clean the things around us, and what would happen if we didn't wash things. Talk about germs and how they can make us sick. Emphasize the importance of washing hands before and after using the bathroom and before eating. What would our dishes and plates look like if we didn't have water? What would our clothes look like? How would we begin to look and smell? Model for your students how to use the washing center. Show them how to wash the toy or dish thoroughly, rinse the soap off and dry it with a towel. Show your students what to do with the objects they have just cleaned. Be sure to discuss with your students how to use the center appropriately. Remind them how to keep the water inside the appropriate tub.

Center Two – Mixing Center

Objective: Students will understand that different substances will change when water is added.

Materials

- water, in an easy-to-pour container
- several small tubs or bowls for mixing
- hand-operated mixers
- spoons
- soap flakes
- blue construction paper
- activity card, laminated (page 155)

Directions: Provide the following for each of your students at the center: a mixing bowl, hand mixer, and soap flakes. (You could also use just one mixer and mixing bowl and have the students take turns mixing the soap flakes.) Demonstrate how the students will fill their bowls halfway with water. Then, gradually add soap flakes, stirring constantly with the mixer, until the mixture becomes a thick, shaving cream consistency. Drop a large dollop of soap mixture onto the blue paper and have the students spread it around with their hands. Kids can make pictures, or you can use it as an opportunity to practice letter, number, or word formation. You may need to add more soap when the mixture thins out.

Note: Parent help will make this center run more smoothly.

Living Things Need Water

Name: _____

Directions: Look at the pictures. Color the pictures that need water. Cross out the items that do not need water.

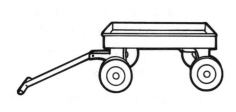

Washing Center

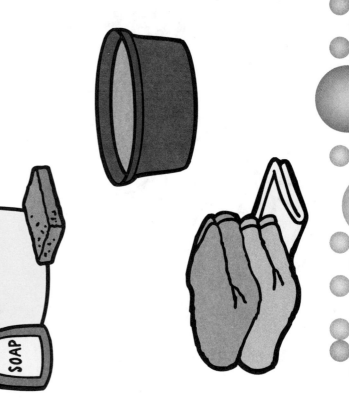

1. Pick an object you would like to wash out of the basket.

2. Gently wash the item using the sponge or rag.

3. Gently place the item in the rinsing water.

4. Dry the item using the towel.

WE WASH OUR HANDS

after playing with pets

before and after eating

after blowing our nose

whenever they are dirty

after using the bathroom

Mixing Center

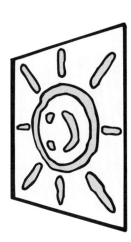

1. Using the measuring cup, fill the mixing bowl halfway with water.

2. Slowly add soap flakes to the water and mix.

3. Keep mixing and adding more soap flakes until the mixture thickens.

4. Make a picture using the soap flake mixture.

 # What Kind of Water?

Name: _____

Directions: Water has many forms. It can be a *solid*, a *liquid*, or a *gas*. Cut out the pictures below and place them in the correct column.

Solid	Liquid	Gas

Physical Science

 # Mixing with Water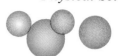

Name: _____

Directions: Mix the ingredients shown on each line. Color the first jar in the row to show how the mixture looked after stirring. Wait a day and recheck the jars. Color the last jar in each row to show the results after one day.

Mixture	After Stirring	Next Day
water + oil =		
water + sand =		
water + flour =		
water + soil =		

 # How I Used Water

Name: _____

Directions: Record how you used water on the lines below.

I used water when I _____

I used water when I _____ .

I used water when I _____ .

I used water when I _____ .

I used water when I _____ .

I used water when I _____ .

Draw a picture of one way in which you used water.

 # Vocabulary Cards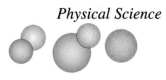

clean	dirty
ice	liquid
melt	solid
wash	water

 # Vocabulary Cards

Water

Water

Water

Water

Water

Water

Water

Water

Tools and Machines

What is a tool? A tool is a hand-held implement used to help accomplish a task.

Where can tools be found? Tools can be found almost anywhere. Simple tools used in the classroom include crayons, erasers, rulers, and pencils. They all help to get a job done. If a student drinks his or her milk using a straw, he or she is using a simple tool—the straw helps the student drink the milk more efficiently. Simple tools that can be found in our homes include things like wooden spoons, forks, spatulas, and even toothbrushes. Rakes, hoes, and shovels are all garden tools. Saws, hammers, and screwdrivers are tools often found in the typical garage or tool shed.

What is a machine? A machine is a tool with more than one part. The parts work together to perform tasks more simply or more quickly. A lever is a simple machine. Other simple machines include the pulley, the wedge, and the wheel and axle. Other machines can be quite complex. Some examples are computers and washing machines. Some machines are small like the calculator and the stapler. Others are quite large like buses and airplanes.

Where can machines be found? Some machines used at school include a clock, tape recorder, computer, and overhead projector. Many homes have eggbeaters, vacuums, hairdryers, and laundry machines.

Why are tools and machines used? Tools and machines help to complete a job more easily or quickly. They cannot work by themselves and must be operated by a human being. Tools and machines may or may not need power, like electricity.

Are tools and machines necessary for survival? Of course not—but they certainly make life easier and more pleasant.

Farmers use many different tools and machines to help them grow food. Doctors use tools and machines to help people get well. Builders use tools and machines to help construct homes, shopping areas, hospitals, etc. Large machines, like cars, trucks, steam shovels, etc., are found in the community. Most machines require electricity or fuel.

It may surprise you to realize that some modern-day conveniences that we take for granted were only developed within the last 100 years—the car (1900), the vacuum cleaner (1901), the airplane (1903), the refrigerator (1911), the television (1927), and the computer (1945).

 # Help Is on the Way!

Name: _____

Directions: Think of a chore or job that you have to do. Create a tool or machine that would help save you time and energy. Draw your invention in the space below. On the lines provided, write about how it would work and what it would do.

_____

_____

_____

_____

_____

Machines and Tools

Objectives

Students will understand the difference between a tool and a machine; that tools and machines are a part of everyday life; that tools and machines make the completion of certain tasks easier and faster.

Materials

- bowl
- box of instant pudding or soap flakes
- eggbeater
- electric mixer
- measuring cups
- milk
- timer
- wire whisk
- "Tools" and "Machines" cards, (pages 171–174)
- Vocabulary Cards, cut and laminated (page 175)

- books and posters about machines and tools
- actual tools and machines
- resealable plastic bags for cards
- copies, "Help Is on the Way!" (page 162)
- copies, "Tools and Machines" (page 168)

Preparation

Gather nonfiction books about tools and machines. Display simple items like small hammers, screwdrivers, pencil sharpeners, safety scissors, staplers, hole punchers, hand stamps, tape dispensers, etc. Find posters of important tool/machine inventions and select interesting biographies of great inventors to share with the class. If available, set out pieces from building toys with moving parts that allow students the opportunity to create their own machines.

Lesson

Discuss with your students the differences between *tools* and *machines*. Ask your students to look around the classroom for examples of tools or machines. Make a list of the examples your students find on the board or on chart paper. Some examples of tools and machines you might find in your classroom: pencils, markers, erasers, rulers, pencil sharpeners, phones, computers, staplers, etc. After you have made a list of the machines and tools you might find in a classroom, discuss and list tools and machines that one might find at home.

Use the worksheet on (page 168) to expand discussions of tools and machines and their uses.

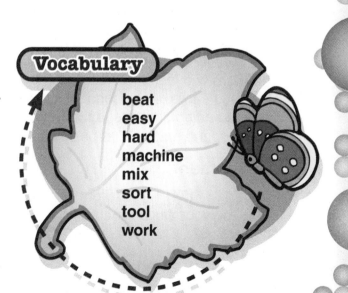

Vocabulary

beat
easy
hard
machine
mix
sort
tool
work

Machines and Tools

Lesson (cont.)

Next, have students think specifically about tools/machines used in preparing a meal. Emphasize to your students that we use machines and tools to help us accomplish a task easier and faster. Inform them that, although machines and tools can help us do work more efficiently, if not used properly, they can also be dangerous. Remind them to make sure an adult is supervising them when they use a machine or a tool.

Have your students gather around a table to make pudding. Remind students about the importance of washing hands before handling or preparing food. Tell your students that you are going to use tools and machines to help you prepare a snack for the class. Begin to prepare the instant pudding, following the directions on the box. When all the ingredients are in the bowl, ask a student to set a timer for one minute. Mix the pudding with the wire whisk (tool). After the minute has passed, have students observe the results. Talk about how the pudding's consistency didn't change "instantly."

Explain that, although tools do help us, sometimes a machine can help us to accomplish something more efficiently. (Remind your students that a machine is an object made up of two or more parts.) Can they think of a machine that might prove more helpful?

Next, try using the eggbeater (machine). Ask a student to set a timer for one minute. After the minute has passed, ask your students to observe the results. Again, talk with your students about how the pudding's consistency still isn't really changing "instantly." Ask them to think of another way to make the pudding firm up faster.

Finally, use an electric mixer (machine). Ask a student to set a timer for one minute. After the minute has passed, ask your students to observe the results. Talk about the tools and machines you used and what worked best. What conclusion can the students draw? **Note**: If you don't want to use food to demonstrate this concept, use soap flakes.

There are two learning centers for this section. The first center involves sorting Tools and Machines cards, the second includes tools and machines and requires adult supervision. Before sending students to the Bubble Beaters Center, model and explain how to use each tool and machine properly.

Assessment

Pages 168, 171–174 challenge students to differentiate between a tool or machine. Stimulate your students' creativity by encouraging them to complete "Help Is on the Way!," on page 162. Can they "invent" their own tool or machine?

Home Conection

Have students choose a room in their homes to survey. Encourage them to list the tools and machines they observed. What is each student's favorite tool or machine?

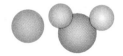 # Tools and Machines

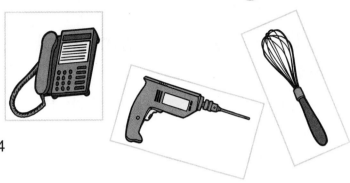

Center One – Sorting

Materials

- "Tools" heading and cards, pages 171–172
- "Machines" heading and cards, page 173–174

Preparation

Arrange the two heading cards on a table in the center. Review the cards with students as a group prior to participation in the center. Stack the "Tools" and "Machine" cards in a pile near the heading cards.

Lesson

Students may work individually or in groups to sort the cards into two groups, one for machines and one for tools. Have students lay the cards under the appropriate title so that all are visible.

Assessment

Students can self-check their ability to distinguish between tools and machines by flipping over the cards. The tool cards all have a "T" on the back and the machine cards are labeled with an "M" on the back of each card.

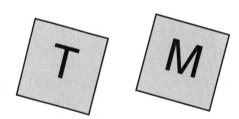

Optional Individual Activities

Make a set of cards for each of your students. The card sets can then be used for a variety of other activities:

- Students can practice placing the cards in alphabetical order.

- Students can sort the cards into different categories—items they have at home/school; items that are bigger/smaller than they are; items that children should not touch/items that are safe for children to use; items that are noisy/quiet; items that do/do not need electricity, etc.

- In pairs, students can use their memorization skills while playing "Concentration" with two sets of cards.

- Students can study the cards and learn to spell all of the words.

- Cut paper to the same size as the cards. Have the students think of other machines and tools, draw them, and add them to their collection. How many can they think of?

Tools and Machines

Center Two – Bubble Beaters

Materials

- four or more bowls
- dishwashing liquid
- eggbeater
- spoons
- whisks
- electric mixer
- measuring cups
- measuring spoons
- towels
- water
- copies, "Bubble Beaters" worksheet, page 170

Lesson

Students will work in groups of four. Each student will get to use one of the items listed above for mixing—spoon, whisk, eggbeater, or hand mixer. Model for your students how to use all the tools properly. Add one tablespoon of dishwashing liquid to each bowl of water. Add more, if necessary, to create more bubbles. Tell your students they are going to choose a tool, carefully place the tool in the water, and try to create bubbles. Can they predict who will have the most bubbles? Tell them to carefully observe and remember the results of everyone in the group.

Another way to do this experiment would be to divide the class into three equal groups. Give each group a different tool or machine with which to beat their dish soap. For example, give one group the wire whisk, one group a hand-operated mixer, and one group the electric mixer. Give each group three minutes to mix the dish soap and discuss the results at the end of the time period.

Safety Note: It is necessary to have an adult supervise this activity.

Assessment

Have students complete the "Bubble Beaters" worksheet on page 170 after working at the center.

Tools and Machines Sort

Tools | Machines

1. Look at the two headings:

2. Take a card from the stack and decide if it is a tool or a machine.

3. Place the card under the correct heading.

4. Sort all the cards.

Tools and Machines

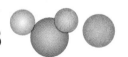

Name: _____

Directions: Draw a line from the object to the appropriate tool or machine.

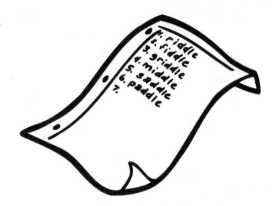

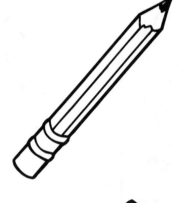

Bubble Beaters

1. Choose a tool or a machine.

2. Add a spoonful of soap to the water.

3. Carefully put the tool or machine in the water.

4. Use the tool or machine to create bubbles.

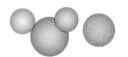

Bubble Beaters

Name: _____

Directions: Cut out the four cards at the bottom of the page. Then, paste the cards in order from the one that made the fewest bubbles (#1) to the one that made the most bubbles (#4).

1	2	3	4

Tools

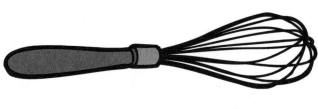

whisk

shovel

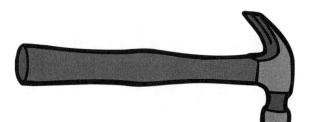

hammer

pencil

screwdriver

spoon

rake

ruler

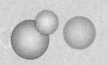

Tools

T   T

T   T

T   T

T   T

Machines

vacuum

whisk

telephone

car

computer

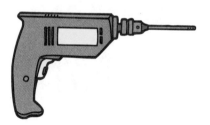
drill

tractor

lawnmower

Machines

M   M

M   M

M   M

M   M

 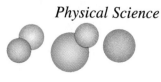
easy	hard
machine	mix
sort	tool
beat	work

Vocabulary Cards

Tools & Machines

Tools & Machines

Tools & Machines

Tools & Machines

Tools & Machines

Tools & Machines

Tools & Machines

Tools & Machines